ENGINEERING PROJECT MANAGEMENT

Also from Blackwell Science:

Managing Risk in Construction Projects
Edited by N. J. Smith
0–632–04243–5

Modern Construction Management
Fourth Edition
Frank Harris and Ronald McCaffer
0–632–03897–7

Risk Management and Construction
Roger Flanagan and George Norman
0–632–02816–5

Project Management in Construction
Third Edition
Anthony Walker
0–632–04071–8

Construction Management: New Directions
Denny McGeorge and Angela Palmer
0–632–04258–3

Corporate Communications in Construction
Christopher N. Preece, Krisen Moodley and Alan Smith
0–632–04906–5

Engineering, Construction and Architectural Management
A journal publishing original research work and technical papers reporting innovative developments in construction and its management.
ISSN: 0969–9988

ENGINEERING PROJECT MANAGEMENT

Edited by

Nigel J. Smith
Professor of Construction Project Management
University of Leeds

b
Blackwell
Science

Editorial Offices:
Osney Mead, Oxford OX2 0EL
25 John Street, London WC1N 2BL
23 Ainslie Place, Edinburgh EH3 6AJ
350 Main Street, Malden
MA 02148 5018, USA
54 University Street, Carlton
Victoria 3053, Australia
10, rue Casimir Delavigne
75006 Paris, France

Other Editorial Offices:

Blackwell Wissenschafts-Verlag GmbH
Kurfürstendamm 57
10707 Berlin, Germany

Blackwell Science KK
MG Kodenmacho Building
7–10 Kodenmacho Nihombashi
Chuo-ku, Tokyo 104, Japan

First published 1995
Reprinted 1996, 1998

Set in 10pt Times
by DP Photosetting, Aylesbury, Bucks
Printed and bound in Great Britain by
MPG Books Ltd, Bodmin, Cornwall

DISTRIBUTORS

Marston Book Services Ltd
PO Box 269
Abingdon
Oxon OX14 4YN
(*Orders:* Tel: 01235 465500
Fax: 01235 465555)

USA
Blackwell Science, Inc.
Commerce Place
350 Main Street
Malden, MA 02148 5018
(*Orders:* Tel: 800 759 6102
781 388 8250
Fax: 781 388 8255)

Canada
Login Brothers Book Company
324 Saulteaux Crescent
Winnipeg, Manitoba R3J 3T2
(*Orders:* Tel: 204 837-2987
Fax: 204 837-3116)

Australia
Blackwell Science Pty Ltd
54 University Street
Carlton, Victoria 3053
(*Orders:* Tel: 03 9347 0300
Fax: 03 9347 5001)

A catalogue record for this title
is available from the British Library

ISBN 0-632-03924-8

Library of Congress
Cataloging-in-Publication Data
is available

For further information on
Blackwell Science, visit our Website:
www.blackwell-science.com

Contents

Preface

In many sectors of industry the significance of good project management in delivering projects in accordance with predetermined objectives has been recognized. Industrialists and the engineering institutions have called for the inclusion of a significant proportion of project management education in higher-level degrees. The Finniston Report in 1980 recognized that project management skills were missing from the complete formation of an engineering education, and this was reinforced when the 1992 White Paper on *Increasing our Competitiveness* commented on the continuing inadequacy.

Many organizations in engineering, finance, business, process and other sectors are appointing people as project managers, some with a very narrow brief and a precise role, whereas others have a more strategic, managerial and multidisciplinary function; this difference is reinforced by a small number of textbooks, each giving a specific viewpoint and interpretation of the role of a project manager. Consequently, there is some confusion over the usage and meaning of the term 'project manager'.

This book provides a clear description of the aim of project management, which is based on the use of good management practice for managing the work of an organization for both its internal and external projects, and hence provides an understanding of the function of the project manager. The authors consider that a project can be any activity that has an identifiable beginning and end and a specific aim or purpose. It can relate to new work, modification, maintenance and refurbishment or decommissioning and demolition. The work can be in any sector and is usually multidisciplinary, but this book concentrates on engineering projects principally in the chemical, civil, electrical and mechanical disciplines.

Project management is in the process of being recognized as a distinct discipline rather than a collection of 'add-on' techniques for cost control and organizational structure. The book draws upon the expertise of the authors to offer explanations of theory and technique combined with practical guidance. It is not a series of check-lists nor does it contain management essays but rather it offers a framework for the application of project management and for the duties and responsibilities of the project manager.

Students of project management studying at higher degree level will find the text and the references beneficial. The book is concerned with the practice and theory of project management, particularly in relation to multidisciplinary engineering projects, both large and small, in the UK and overseas. Its thorough treatment of the practical issues of investment appraisal and risk will also assist the recently appointed project manager.

Acknowledgements

I am particularly grateful to my co-authors and fellow contributors to this book, and would like to thank Dr Denise Bower, Dr Tony Merna, Professor Peter Thompson, Professor Stephen Wearne, and Mr Ian Vickridge. I am also grateful to the postgraduate students of UMIST who provided valuable critical assessments and comments throughout their respective project management lecture courses.

I would also like to thank all the staff of the Project Management Group, at UMIST – past and present. Many former members of the group have made significant contributions to the development of the understanding of project management, and I would like to acknowledge individually Dr Martin Barnes, Mr Ross Hayes, Mr Tom Nicholson, Mr Paul Jobling and Professor John Perry, who each influenced the style and structure of multidisciplinary project management teaching at UMIST.

The editor and authors would like to express their appreciation for Thomas Ng, Luiz Carlos Pinta da Silva Filho and Kareem Yusuf of the Project Management Group, who jointly prepared the artwork.

I would especially like to thank Mrs Joan Carey for processing, checking and improving each of the many draft versions of every chapter. Nevertheless, the responsibility for any errors remains entirely my own.

In addition to all those acknowledged above, I would like to express my gratitude to all those who assisted in the preparation of this new print run. In particular I would like to mention Luiz Carlos Pinta da Silva Filho of the Project Management Group, UMIST, and Kareem Yusuf, now of the Construction Management Group, University of Leeds, for their skills in revising the artwork. I would also like to thank Judi Ogden for all the word processing and administrative support work.

N.J. Smith

List of Contributors

University of Leeds Construction Management Group

Professor Nigel Smith is Head of the Construction Management Group in the Department of Civil Engineering, University of Leeds. From a small base the group has expanded significantly in recent years and is now particularly active in research and consultancy concerning project management for the Construction Industry, both in the UK and overseas.

Research funding has been attributed from a range of national and international funding bodies and from organisations within the construction industry itself. Projects completed include the provision of management information systems, work which has placed increased emphasis on the role of Information Technology in project management. The group has been appointed to study competitiveness in conjunction with six European airports and to review the practicability of the implementation of the Latham Report within the National Health Service.

Nigel J. Smith BSc, MSc, PhD, CEng, MICE, MAPM is Professor of Construction Project Management, Department of Civil Engineering, University of Leeds and formerly a member of the Project Management Group, Department of Civil and Structural Engineering, UMIST. He is a Project Director of the European Construction Institute. After graduating in Civil Engineering at Birmingham University, he gained industrial experience with Wimpey, North East Road Construction Unit and the Department of Transport. In 1991 he was Chairman of a Joint Meeting of the International Association of Cost Engineers and the International Association of Project Managers on Estimating in Trondheim. He is author of many publications relating to project management for construction. He is also the Examiner in Project Management for the Diploma in Engineering Management, a qualification jointly established by the Institution of Chemical Engineers, the Institution of Civil Engineers, the Institution of Electrical Engineers, the Institution of Measurement and Control, the Institution of Mechanical Engineers and the Institution of Structural Engineers.

UMIST Project Management Group

This academic group, based at the University of Manchester, Institute of Science and Technology (UMIST), was founded in 1965 with the aim of contributing to improved performance in engineering project management. The group is active in research and consultancy in many parts of the world, and is recognized internationally as a centre of excellence and innovation in multidisciplinary project management.

Studies of specific topics included risk analyses for major projects such as the Severn tidal power scheme and involvement in developing the contracts and control systems for specific projects in Third World countries and observing the results. The conclusions of these studies were increasingly directed towards defining the lessons for project promoters and the choices in project and contract strategies. The group was also commissioned to recommend policies on cost estimating and contract strategy for overseas projects, to observe and report on other projects, and to review the 'state-of-the-art' in the offshore and onshore construction industries.

The resulting body of knowledge, which extends to all aspects of engineering project management, forms the basis of the teaching material presented in this book.

Denise Bower BEng, PhD, AMICE is a lecturer in the Project Management Group at UMIST, a position partly funded by Shell. She graduated in civil engineering, worked with a contractor on site, and then moved into her current research post working on contract payment systems in civil engineering. Her recent work includes evaluating computer software for project management and earned value techniques for MoD contracts, case studies of the management of major projects on an operating production site, assessment of a process facility built under a concession contract, and recommendations of contract strategies for overseas projects. She is the joint author of several case studies of the management of recent construction projects.

Tony Merna BA, MPhil, PhD, CEng, MICE, MAPM, MIQA is senior partner of the Oriel Group Practice (UK). He spent 12 years managing multidisciplinary projects in Africa and the Middle East. Returning to the UK in 1987 he was awarded an MPhil for research into turnkey contracts before going overseas again to work on a major infrastructure project in Jakarta. In 1990 he returned to academic research at UMIST as a research associate investigating the technical, legal and commercial aspects of concession contracts. Having gained his doctorate, Tony founded Oriel Group Practice, an independent consultancy specializing in services dealing with project management, risk management and privately financed concession contracts. Tony is based in Manchester, where he is involved in the project management of a number of projects utilizing private finance and general project management.

Peter Thompson BSc (Eng), MSc, CEng, FICE, MIWEM has recently returned as the AMEC Professor of Engineering Project Management and Director of the Project Management Group at UMIST. After graduating at Queen Mary College, he gained extensive industrial experience on power stations, oil installations and major water supply projects. Since moving to UMIST he has specialized in the financial aspects of contracts and projects, project appraisal and contract strategy with emphasis on promoter project management. He has lectured extensively in the UK and abroad and acted as a consultant on many overseas projects. Recent work includes a study of partnering and alliance contracts in the process and offshore industries and further research on risk management. He is currently studying the changing role of the promoter related to different contractual arrangements.

Ian Vickridge BSc, MSc, CEng, MICE, MIWEM is a senior lecturer in the Department of Civil and Structural Engineering, UMIST. After graduating from Strathclyde, his initial employment was with consulting engineers Sir William Halcrow and then John Taylor & Sons, before his first academic appointments at the Polytechnic of Wales and Hong Kong Polytechnic. He then returned to industry to work on water treatment facilities in the UK with Nicholas O'Dwyer and Partners and in Saudi Arabia with Biwater Treatment. He then spent four years at the Nanyang Technological Institute in Singapore as a senior lecturer before joining UMIST in 1984. He has specialized in water engineering, project management and civil engineering in developing countries. Currently he is Executive Secretary and Director of the United Kingdom Society for Trenchless Technology.

Stephen Wearne BSc (Eng), PhD, DIC, CEng, FICE, FIMechE, FRSA, MAPM is a consultant and a senior research fellow in the UMIST

Project Management Group. His current work with the group is on institutional risks of projects, joint ventures and consortia. After his initial training in mechanical and water power engineering he was employed on the construction, design, planning and coordination of projects in Spain, Scotland and South America. In 1957 he joined turnkey contractors on construction and design, and then contract and project management of nuclear power projects in this country and Japan. He moved into research and teaching in engineering management in 1964, first at UMIST, and then from 1973 to 1983 as Professor of Technological Management at the University of Bradford. His research has included studies of design and project teams, engineering contracts, project control, responsibilities for plant commissioning, and the managerial tasks of engineers in their careers. He was first chairman of the UK Engineering Project Management Forum initiated in 1985 by the National Economic Development Council and the Institutions of Civil, Mechanical, Electrical and Chemical Engineers.

List of Abbreviations

The following abbreviations are used in publications on British and some international engineering and construction contracts.

ACE	Association of Consulting Engineers.
BAQ	Bills of approximate quantities.
BEAMA	British Electrotechnical and Allied Manufacturers Association.
BEC	Building Employers' Confederation.
BOOT	Build-own-operate-transfer project.
BoQ	Bills of quantities.
BOT	Build-operate-transfer project.
BS	British Standard.
CCPI	Committee for Coordinated Project Information (RIBA, RICS, BEC and ACE).
CCSJC	Conditions of Contract Standing Joint Committee (ICE, ACE and FCEC).
CESMM	Civil Engineering Standard Method of Measurement (for a BoQ).
CIArb	Chartered Institute of Arbitrators.
CIB	International Council for Building Research Studies & Documentation.
CIC	Construction Industry Council.
CII	Construction Industry Institute (Texas).
CIPFA	Chartered Institute of Public Finance and Accountancy.
CIRIA	Construction Industry Research and Information Association.
CPA	Contract price adjustment of tendered rates or prices to allow or partly allow for escalation (changes in material, labour or other costs during the work under a contract).*

CTR	Cost-time-resource basis of payment.
EC	The European Community.
ECE	United Nations Economic Commission for Europe.
ECI	European Construction Institute.
EDF	European Development Fund.
EEF	Engineering Employers Federation.
EMC	Engineer and manage contract.
EPC	Engineer-procure-construct contract.
EPCm	Engineer-procure-manage construction contract.
EPIC	Engineer-procure-install-commission contract.
EU	European Union.
FASSC	Federation of Associations of Specialists and Sub-Contractors.
FCEC	Federation of Civil Engineering Contractors
FIDIC	Fédération Internationale des Ingénieurs Conseils (Lausanne).
ICC	International Chamber of Commerce.
ICE	Institution of Civil Engineers.
IChemE	Institution of Chemical Engineers.
IEE	Institution of Electrical Engineers.
IMechE	Institution of Mechanical Engineers.
ITT	Invitation to tender.
JCT	Joint Contracts Tribunal for the Standard Building Contract (RIBA and other building industry organizations).
NAECI	National Agreement for the Engineering Construction Industry.
NECEA	National Engineering Contractors Employers Association.
NEDO	National Economic Development Office.
NJCECI	National Joint Council for the Engineering Construction Industry.
OED	*Oxford English Dictionary.*
OMT	Operate-maintain-train contract.
PCA	Price cost adjustment, as CPA.
PIT	Progressing, inspection and testing.
PSC	Project services contract.
QA	Quality assurance.
QS	Quantity surveyor.
RIBA	Royal Institute of British Architects.
RICS	Royal Institution of Chartered Surveyors.
SI	Site instruction.†

SMM	Standard method of measurement (for measuring work when using a BoQ).
SMMI&EC	Standard Method of Measurement for Industrial and Engineering Construction.
SO	Supervising or Superintending Officer.
TQ	Technical query.
TQM	Total quality management.
VO	Variation order.
WBS	Work breakdown structure.

* CPA is also used in project planning to mean critical path analysis.

† SI ought to be used only to mean the Système International d'Unités (International System of Units).

Part I
Project Concepts

Chapter 1
Projects and Project Management

This chapter introduces the process of project initiation and concludes by discussing the underlying causes of success in projects now and in the future. To facilitate an understanding of the nature of 'projects' and the role of 'project management' it is first necessary to confirm the accepted meaning of these terms. This usage is consistent throughout the book and should be borne in mind when reading other texts in which differing meanings have been adopted.

1.1 Projects

A project can be any new structure, plant, process, system or software, large or small, or the replacement, refurbishing, renewal or removal of an existing one. It is a one-off investment.

One project may be much the same as a previous one and different from it only in detail to suit a change in market or a new site. The differences may extend to some novelty in the product, in the system of production, or in the equipment and structures forming a system. Every new design of car, aircraft, ship, refrigerator, computer, crane, steel mill, refinery, production line, sewer, road, bridge, dock, dam, power station, control system, building or software package is a project. So are many smaller examples, and a package of work for any such project can in turn be a subsidiary project.

Projects thus vary in scale and complexity from small improvements to products to large capital investments. The common use of the word 'project' for all of them is logical because every one is:

- an investment of resources for an objective;
- a cause of irreversible change.

A project is an investment of resources to produce goods or services. In other words it costs money. The normal criterion for investing in a proposed project is therefore that the goods or services produced are more valuable than the predicted cost of the project.

To get value from the investment, a project usually has a defined date for completion. As a result, the work for a project is a period of intense engineering and other activities but short in its duration relative to the subsequent working life of the investment.

A *programme of projects* means a set of related projects.

A *project programme* means a list of activities for a project showing the dates for starting and finishing them. It is also called a *schedule*.

Program in the UK usually means computer software, and this is how it is used in this book. Program in the USA can mean any of the above.

1.2 Project management

> Project management is the planning, organization, monitoring and control of all aspects of a project and the motivation of all involved to achieve project objectives safely and within a defined time, cost and performance (Association of Project Managers, 1995).

Projects are the basis of the success of all production and services. In most if not all organizations survival and success depend upon a range of projects varying from the large and novel to minor alterations of existing investments.

Project management is needed to look ahead at the needs and risks, communicate the plans and priorities, anticipate problems, assess progress and trends, get quality and value for money, and change the plans if needed to achieve objectives.

The needs for project management are dependent upon the relative size, complexity, urgency, importance and novelty of a project. The needs are also greater where projects are interdependent, particularly those competing for the same resources.

Every project disturbs the status quo, if only a little. This inherent characteristic of engineering does not fit with the bureaucratic administrative convenience that decisions are precedents that establish rules for solving a category of problem. Projects only establish the precedent that categories of problems will change. The older the industry the lesser may be its apparent rate of change, but innovation displacing adaptation is the general trend.

Every project proceeds through a cycle of activities, starting with the initial study of ideas about what may be needed and how it might be achieved, and ending with a much more intense concentration of resources to complete it.

Stages of work as indicated in Figure 1.1 are therefore typical of engineering projects. Each stage marks a change in the nature, complexity and speed of the activities and the resources employed as a project proceeds.

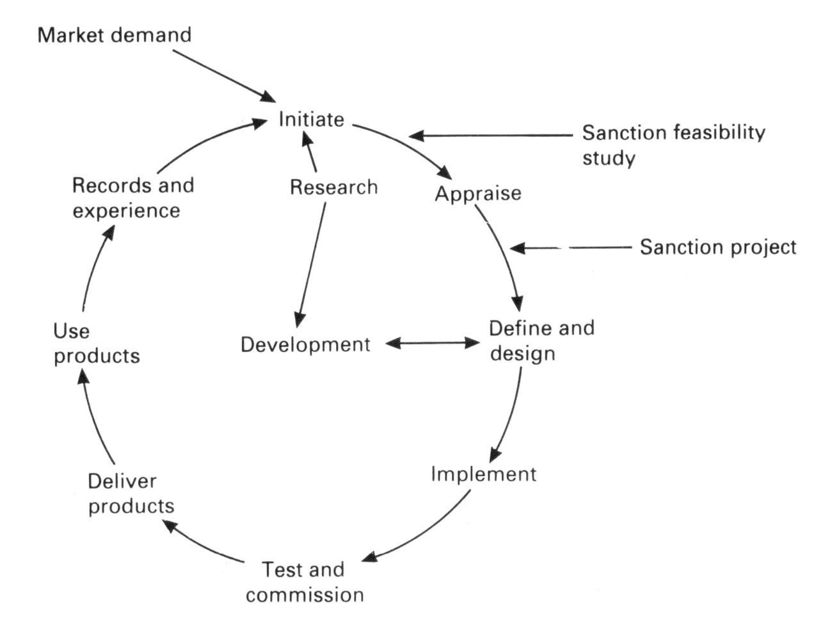

Figure 1.1 Cycle of stages of work for a project.

The durations of the stages vary from project to project, with sometimes delay between one and the next. They can also overlap. Figure 1.1 shows the sequence of starting them. It is not meant to show that one must be completed before the next is started. The objective of the sequence should be to produce a useful result, so that the purpose of each stage should be to enable the next to proceed.

1.3 Project initiation

As shown in Figure 1.1, a project is likely to be initiated when whoever is its promoter predicts that there will be a demand for the goods or services that the project might produce. The ideas for the project should then draw on records and experience of previous projects and the results from research indicating new possibilities. These three sources of information ought to be brought together at this stage of a project.

Usually at this stage there will be alternative ideas or schemes that seem likely to meet the demand. Proceeding further requires the promoter to authorize the use of some resources to investigate these ideas and the potential demand for the project. The term *sanction* is used here to mean the decision to incur the cost of these investigations. The cycle then proceeds to appraise the ideas to compare their predicted cost with predicted value.

In emergencies, appraisal is omitted, or if a project is urgent no time may be used in trying to optimize the proposal. More commonly, alternative proposals have to be evaluated in order to decide whether to proceed and how best to do so in order to achieve the promoter's objectives.

Feasibility study

The appraisal stage is also known as the *feasibility study*.

Its results can only be probabilistic, as they are based upon predictions of the demand and of costs whose reliability varies according to the quality of the information used, the novelty of the proposals, and the amount and quality of the resources available to investigate the risks that could affect the project and its useful life.

Repetition of the work up to this point is often needed after the first appraisal, as its results may show that better information is needed on the possible demand, or the conclusions of the appraisal are disappointing and revised ideas are needed that are more likely to meet a demand. More expenditure has to be sanctioned to do so. Repetition of the work may also be needed because the information used to predict the demand for the project has changed during this time. Feasibility studies may therefore have to be repeated several times.

Concluding this work may take time. Its result is quite specific: the sanctioning or the rejection of a proposed project. If a project is selected the activities change from assessing whether it should proceed to deciding how best it should be realized and to specifying *what* needs to be done.

Design, development and research

Design ideas are usually the start of possible projects, and alternatives investigated before estimating costs and evaluating whether to proceed any further. The main design stage of deciding how to use materials to realize projects usually follows evaluation and selection, as indicated in Figure 1.1. The decisions made in design determine almost entirely the quality and cost and therefore the success of a project. Scale and specialization increase rapidly as it proceeds.

Development in the cycle is experimental and analytical work to test means of achieving a predicted performance. Research ascertains properties and potential performance. The two are distinct in their objectives. Design and development share one objective, that of making ideas succeed. Their relationship is therefore important, as indicated in Figure 1.1.

Most of the design and supporting development work for a project usually follows the decision to proceed. They may be taken in substages so as to investigate novel problems and review predictions of cost and value before continuing with a greater investment of resources.

Project implementation

There follows the largest scale of activities and the variety of physical work to implement a project, particularly the manufacture of equipment for it and construction work.

Most companies and public bodies who promote new capital expenditure projects employ contractors and subcontractors from this stage on to supply equipment or carry out construction. For internal projects within firms there is the equivalent internal process of placing orders to authorize expenditure on labour and materials.

Sections of a project can proceed at different speeds in design and consequent stages, but all must come together to test and commission the resulting facility. The project has then reached its productive stage. It should then be meeting the specified objectives of the project.

The problems in meeting objectives vary from project to project. They vary in content and in the extent to which experience can be adapted from previous projects in order to avoid novel problems. The criteria for appraisal also vary from industry to industry, but common to all projects is the need to achieve a sequence of decisions and activities as indicated in Figure 1.1.

Figure 1.1 is a model of what may be typical of the sequence of work for one project. Projects are rarely carried out in isolation from others. At the start, alternative projects may be under consideration, and in the appraisal stage competing for selection. Those selected are then likely to share design resources with others that may be otherwise unconnected, because of the potential advantages of sharing expertise and other resources, but will therefore be in competition with them for the use of these resources. Similarly through all the subsequent stages in the cycle.

A project is thus likely to be cross-linked with others at every stage shown in Figure 1.1. These links enable people and firms to specialise in a stage or subpart of the work for many projects. The consequence is often that any one project depends upon the work of several departments or firms, each of which is likely to be engaged on a variety of projects for a variety of customers. In all of these organizations there may therefore be conflicts in utilizing resources to meet the competing needs of a number of projects, and each promoter investing in a project may have problems in achieving the sequence of activities that best suits his interests.

1.4 Project success

The evidence indicates that success of projects now and in the future may depend upon the following.

Definition of project objectives

The greatest lesson of project management is that the first task is to establish, define and communicate clear objectives for every project.

Risks

To succeed, the promoter's team should then assess the uncertainties of meeting the project objectives. If the risks are not identified, success cannot be achieved. The volume of events taking the team by surprise will be just too great for them to have any chance of meeting the objectives.

Early decisions

Many project successes demonstrate the value of completing much of the design and agreeing a project execution plan before commitment to the costly work of manufacturing hardware or constructing things on a site.

Project planning

The form and the amount of planning has to be just right. Not enough and a project is doomed to collapse from the unexpected. Too detailed plans will quickly become out of date and will be ignored.

Time and money

Planning when to do work and estimating what the resources required for it will cost must be considered together, except in emergencies.

Emergencies and urgency

A project is urgent if the value of completing it faster than normal is greater than the extra cost of the faster working. The designation of 'emergency' should be limited to work where the cost is no restraint on using any resources to work as fast as physically possible, for instance rescue operations to save life. An emergency is rare.

A committed project team

Dispersed project teams correlate with failure; concentration correlates with success. The committed project team should be located where the main risks have to be managed.

Separation of people causes misunderstanding of objectives, communication errors and poor use of expertise and ideas. The people contributing part-time or full-time to a project should feel that they are committed to a team.

The team should be assembled in time to assess and plan their work and their system of communications. Consultants, suppliers, contractors and others who are to provide goods and services should likewise be appointed in time to mobilize their resources, train and brief staff and assess and plan their work.

Representation in decisions

Success requires the downstream parties to be involved in deciding how to achieve the objectives of projects and, sometimes, in setting the objectives themselves. Human systems do not work well if the people who make the initial decisions do not involve those who will be affected later.

Communications

The nature of the work for a project changes month by month. So do the communications needed. The volume and importance of communications can be huge.

Many projects have failed because communications were poorly organized. A system of communication needs to be planned and monitored; otherwise information is too late or goes to the wrong place for decisions and becomes mere records, of little value to control. The records then concentrate on allocation of blame for problems rather than on stimulating decisions to control them. The results of informal communications also need to be known and corrected, as bad news often travels inaccurately.

Promoter and the Leader

Every project large or small needs a real promoter, a project champion who is committed to its success.

Power over the resources needed to deliver a project must be given to one person who is expected to use it to avoid as well as to manage problems. In the rest of this book we call this person the *project manager*. It may not be a separate job, depending upon the size and remoteness of the project.

In turn, every subproject should have its leader with power over the resources it needs.

Delegation of authority

Inadequate delegation of authority has caused the failure of many projects, particularly where decisions have been restrained by requirements for approval by people remote from a problem that have delayed actions and so caused crises, extra costs and loss of respect and confidence in management.

Many projects have failed because authority for parts of a large project was delegated to people who did not have the ability and experience to make the decisions delegated to them. Good delegation requires prior checking that the recipients of delegated authority are equipped to make the decisions delegated to them and then monitoring how effectively they are making their decisions. This does not mean making their decisions for them.

Changes to responsibilities, project scope and plans

Some crises and resulting quick changes to plans are unavoidable during many projects. Drive in solving problems is then very valuable, but failure to think through the decisions about problems can cause greater problems and loss of confidence in project leaders.

Control

If the plan for a project is good, the circumstances it assumes materialize and the plan is well communicated, few control decisions and actions are required. Much more is needed if circumstances do change or people do not know the plan or understand and accept it.

Control is no substitute for planning. It can waste potentially productive time in reporting and explaining events too late to influence them.

Reasons for decisions

In project management, every decision leads to the next one and depends upon the one before. The reasons for decisions have to be understood above, below, before and after in order to guide the subsequent decisions. Without this, divergence from the objectives is almost inevitable – and failure is its other name.

Failure to give reasons with decisions and to check that they are accurately understood by their recipients can cause divergent and inconsistent actions. Skill and patience in communication are particularly needed in the rapidly changing relationships typical of the final stages of large projects.

Using past experience

Success is more likely if technical and project experience from previous projects is drawn upon deliberately and from wherever it is available. Perhaps the frequent failure to do this is another consequence of projects appearing to be unique. It is often easier to say 'this one is different' than to take the trouble to draw experience from the ones before. All projects have similarities and differences. The ability to transfer experience forward by making the comparisons is one of the hallmarks of a mature applied science.

Contract strategy

Contract terms should be designed to motivate all parties to try to achieve the objectives of the project and to provide a basis for project management.

Contract responsibilities and communications must be clear, and not antagonistic. The terms of contracts should allocate the risks appropriately between customers, suppliers, contractors and subcontractors.

Adapting to external changes

Market conditions, customer's wishes and other circumstances change and technical problems appear as a project proceeds. Project managers have to be adaptable to these changes yet prepared to deter the avoidable ones.

Induction, team building and counselling

Success in projects requires people to be brought into a team effectively and rapidly using a deliberate process of induction. Success requires teamwork to be developed and sustained professionally. It requires people to counsel each other across levels of the organization, to review performance, to improve, to move sideways when circumstances require, and to respond to difficulties.

Training

Project management demands intelligence, judgement, energy and persistence. Training cannot create these qualities or substitute for them, but it can greatly help people to learn from their own and other people's experience.

A completed large project can require retraining of general management to understand and obtain full benefit from its effect on corporate operations.

Towards perfect projects

The chapters of this book describe the techniques and systems that can be used to apply these lessons of experience. All of them should be considered, but some chosen as priorities depending upon a situation and its problems.

All improvements cost effort and money. Cost is often given as a reason not to make a change. If so, the organization should also estimate the cost of not removing a problem.

Further reading

Baker, B.N. (1988) *Lessons Learned from a Variety of Project Failures*, Proceedings of the 9th INTERNET World Congress on Project Management, Glasgow, **1**, pp. 113–118.

Barnes, N.M.L. and Wearne, S.H. (1993) *The Future for Major Project Management*, Proceedings of the 1st British Project Management Colloquium, Henley.

Construction Industry Institute (1990) *Potential for Construction Industry Improvement*, Source Document SD-62, Construction Industry Institute, Texas.

de Wit, A. (1988) 'Measurement of project success', *International Journal of Project Management*, **6**(3), pp. 164–170.

Internet World Congress (1992) *Project Management without Boundaries*, Proceedings of the 11th Internet World Congress, Florence.

Iyer, A. and Thomason, D. (1991) *An Empirical Investigation ... to Define the Variables Most Prevalent in Project Successes and Failures*, Proceedings of the Project Management Institute Seminar/Symposium, Dallas, pp. 522–527.

Merrow, E.W., with McDonnell, L. and Argüden, R.Y. (1988) *Understanding the Outcomes of Megaprojects: A Quantitative Analysis of Very Large Civilian Projects*, Rand Corporation, Santa Monica.

Morris, P.W.G. (1993) *The Management of Projects*, Thomas Telford, London.

Morris, P.W.G. and Hough, G.H. (1987) *The Anatomy of Major Projects*, Wiley, Chichester.

O'Connell, F. (1994) *How to Run Successful Projects*, Prentice-Hall, Hemel Hempstead.

Thompson, P.A. and Perry, J.G., eds (1992) *Engineering Construction Risks – A Guide to project Risk Analysis and Risk Management*, Thomas Telford, London.

Wearne, S.H. (1995) 'Towards a science of project management', *Project*, July, pp. 6–7.

Wearne, S.H. *et al.*, (1989) *Control of Engineering Projects*, 2nd edn, Thomas Telford, London.

Chapter 2

Project Appraisal and Risk Management

Project appraisal or feasibility study is an important stage in the evolution of a project. It is important to consider alternatives, identify and assess risks, at a time when data are uncertain or unavailable. This chapter outlines the stages of a project and describes in detail risk management techniques. The text does not concentrate on mathematical analysis but rather on the purpose and objectives of using a logical process of managing project risk.

2.1 Investment

The individual project, however significant and potentially beneficial to the promoting organization, will only constitute part of corporate business. It is also likely that, in the early stages of the project cycle, several alternative projects will be competing for available resources, particularly finance. The progress of any project will therefore be subject to investment decisions by the parent organization that may allow the project to proceed.

In most engineering projects the rate of expenditure changes dramatically as the project moves from the early stages of studies and evaluations, which consume mainly human expertise and analytical skills, to design, manufacture and construction of a physical facility. A typical investment curve is shown in Figure 2.1, which indicates that considerable cost will be incurred before any benefit accrues to the promoter from use of the completed project (or asset).

When considering the investment curve, the life cycle of the project splits into three major phases: appraisal and implementation of the project followed by operation of the completed facility. The precise shape of the curve will be influenced by the nature of the project, by external factors such as statutory approvals, and also by the project objectives. In the public sector, appraisal may extend over many years

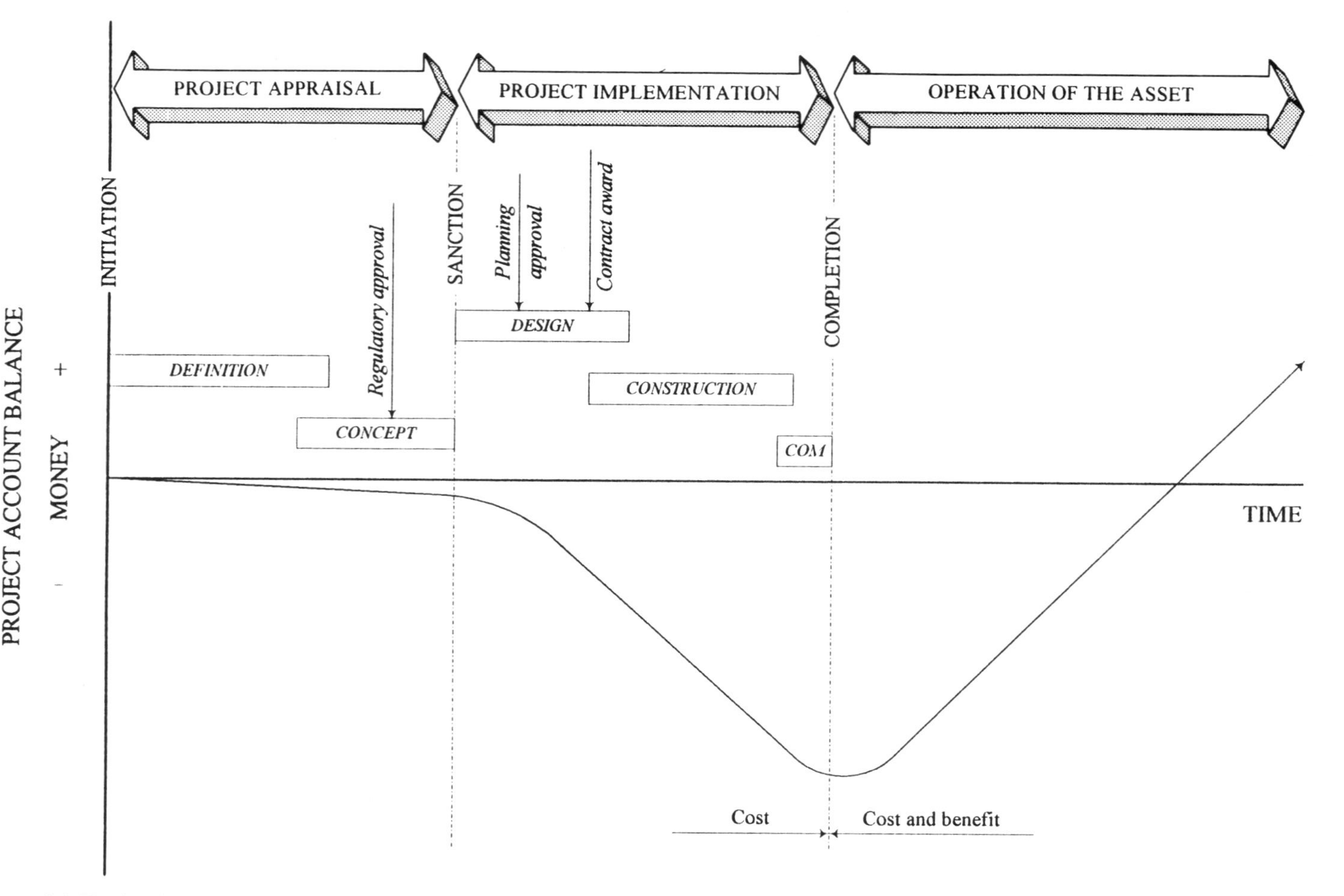

Figure 2.1 Typical investment curve. The period of operation has been truncated on this diagram. In reality one would expect the project to generate significant benefit as illustrated in Figure 7.5 (p.114).

and be subjected to several intermediate decisions to proceed. In a commercial situation the need for early entry into a competitive market may outweigh all other considerations.

Two important factors emerge when studying the investment curve:

- that interest payments compounded over the entire period when the project account is in deficit (below the axis in Figure 2.1) will form a significant element of project cost;
- that the investor will not derive any benefit until the project is completed and is in use.

It is likely that at least two independent and formal investment decisions will be necessary – labelled 'initiation' and 'sanction' on this diagram. The first signifies that the ill-defined idea evolved from research and studies of demand is perceived to offer sufficient potential benefit to warrant the allocation of a specific project budget for further studies and development of the project concept. The subsequent sanction decision signifies acceptance or rejection of these detailed proposals. If positive, the organization will then proceed with the major part of the investment in the expectation of deriving some predicted benefit when the project is completed.

All these estimates and predictions, which frequently extend over periods of many years, will generate different degrees of uncertainty. The sanction decision therefore implies that the investor is prepared to take risk.

2.2 Sanction

When the project is sanctioned the investing organization is committing itself to major expenditure and is assuming the associated risks. This is the key decision in the life cycle of the project. In order to make a well-researched decision the promoter will require:

- *Clear objectives.* The promoter's objectives in pursuing this investment must be clearly stated and agreed by senior management early in the appraisal phase, for all that follows is directed at achievement of these objectives in the most effective manner. The primary objectives of quality, time and cost may well conflict, and it is particularly important that the project team know whether minimum time for completion or minimum cost is the priority. These are rarely compatible, and this requirement will greatly influence both appraisal and implementation of the project.

- *Market intelligence.* This relates to the commercial environment in which the project will be developed and later operated. It is necessary to study and predict trends in the market and the economy, anticipate technological developments and the actions of competitors.
- *Realistic estimates/predictions.* It is easy to be over-optimistic when promoting a new project. Estimates and predictions made during appraisal will extend over the whole life cycle of implementation and operation of the project. Consequently, single-figure estimates are likely to be misleading, and due allowance for uncertainty and exclusions should be included.
- *Assessment of risk.* A thorough study of the uncertainties associated with the investment will help to establish confidence in the estimate and allocate appropriate contingencies. More importantly at this early stage of project development it will highlight areas where more information is needed, and frequently generate imaginative responses to potential problems, thereby reducing risk.
- *Project execution plan.* This should give guidance on the most effective way to implement the project and to achieve the project objectives, taking account of all constraints and risks. Ideally this plan will define the likely contract strategy, and include a programme showing the timing of key decisions and award of contracts.

It is widely held that the success of the venture is greatly dependent on the effort expended during the appraisal preceding sanction. There is, however, conflict between the desire to gain more information and thereby reduce uncertainty, the need to minimize the period of investment, and the knowledge that expenditure on appraisal will have to be written off if the project is not sanctioned.

Expenditure on appraisal of major engineering projects rarely exceeds 10% of the capital cost of the project. The outcome of the appraisal as defined in the concept and brief accepted at sanction will, however, freeze 80% of the cost. The opportunity to reduce cost during the subsequent implementation phase is relatively small, as shown in Figure 2.2.

2.3 Project appraisal and selection

Project appraisal is a process of investigation, review, and evaluation undertaken as the project or alternative concepts of the project are defined. This study is designed to assist the promoter to reach informed and rational choices concerning the nature and scale of investment in the project and to provide the brief for subsequent implementation. The core

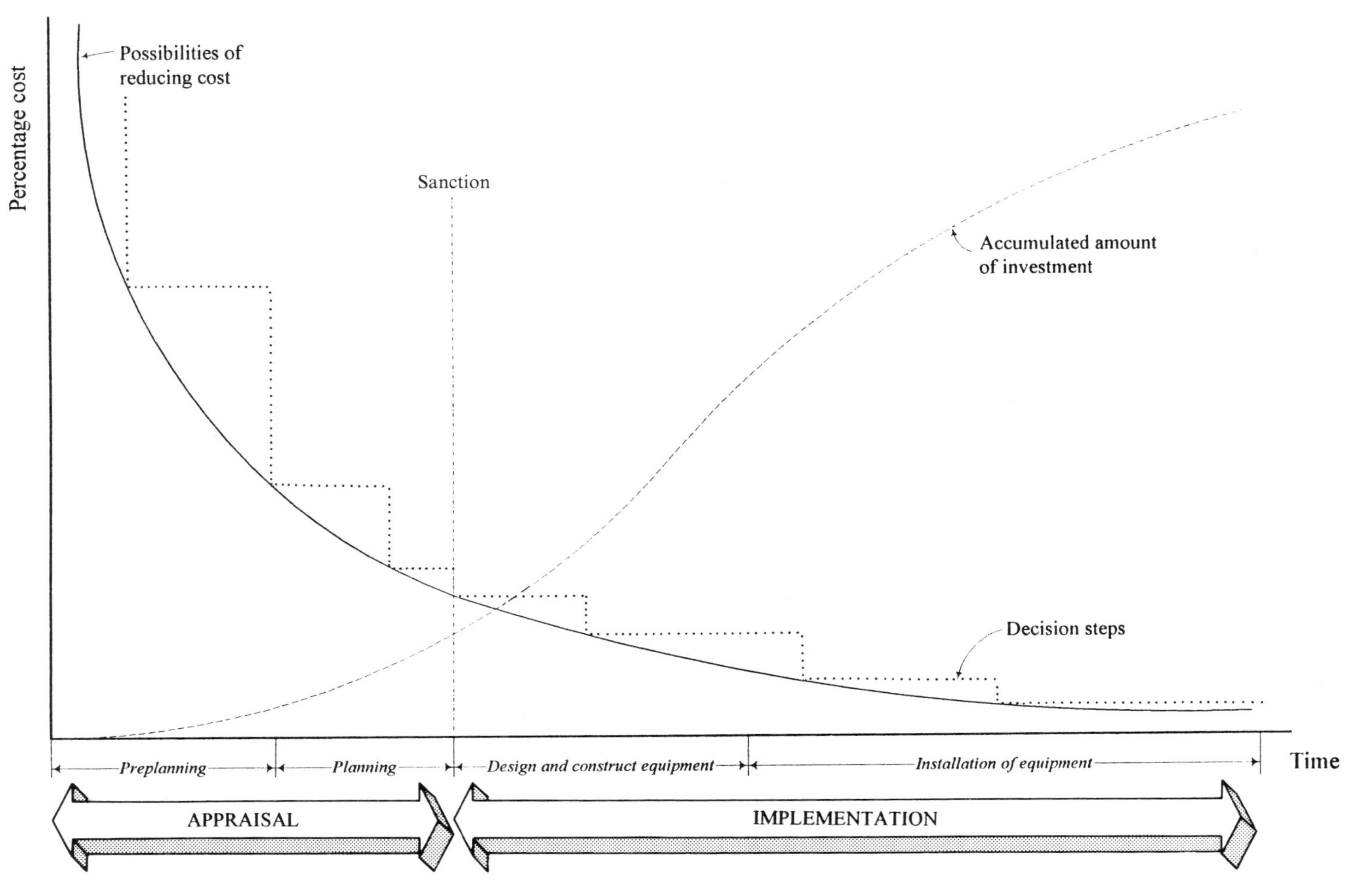

Figure 2.2 Change in the cost of decision making with time.

of the process is an economic evaluation, based on a cash flow analysis of all costs and benefits that can be valued in money terms, which contributes to a broader assessment called cost–benefit analysis. A feasibility study may form part of the appraisal.

Appraisal is likely to be a cyclic process repeated as new ideas are developed, additional information is received, and uncertainty is reduced, until the promoter is able to make the critical decision to sanction implementation of the project and commit the investment in anticipation of the predicted return.

It is important to realize that, if the results of the appraisal are unfavourable, this is the time to defer further work or abandon the project. The consequences of inadequate or unrealistic appraisal can be expensive – as in the case of the Montreal Olympics stadium – or disastrous.

Ideally all alternative concepts and ways of achieving the project objectives should be considered. The resulting proposal prepared for sanction must define the major parameters of the project: the location, the technology to be used, the size of the facility, the sources of finance and raw materials, together with forecasts of the market and the predictions of the cost–benefit of the investment. There is usually an alternative way to utilize resources, especially money, and this is capable of being quantified, however roughly.

Investment decisions may be constrained by non-monetary factors such as:

- organizational policy, strategy and objectives;
- availability of resources such as manpower, management, or technology.

Programme

It will be necessary to decide when is the best time to start the project based on the previous considerations. Normally this means as soon as possible, because no profit can be made until the project is completed. Indeed, it may be that market conditions or other commitments impose a programme deadline, i.e. a customer will not buy your product unless he can have it in two years' time, when his processing factory will be ready. In inflationary times, it is doubly important to complete a project as soon as possible because of the adverse relationship between time and money. The cost of a project will double in 7.25 years at a rate of inflation of 10%.

It will therefore be necessary to determine the duration of the appraisal design and construction phases:

(1) so that the operation date can be determined;
(2) so that project costs can be determined; and
(3) so that the promoter's liabilities can be assessed and checked for viability. It may well be that the promoter's cash availability defines the speed at which the project can proceed.

The importance of time should be recognized throughout the appraisal. Many costs are time-related and would be extended by any delay. The programme must therefore be realistic and its significance taken fully into account when determining the project objectives.

Risk and uncertainty

The greatest degree of uncertainty about the future is encountered early in the life of a new project. Decisions taken during the appraisal stage have a very large impact on final cost, duration and benefits. The extent and effects of change are frequently underestimated during this phase although these are often considerable, particularly in developing countries and remote locations. The overriding conclusion drawn from recent research is that all parties involved in construction projects would benefit greatly from reductions in uncertainty prior to financial commitment.

At the appraisal stage the engineering and project management input will normally concentrate on providing:

- realistic estimate of capital and running costs;
- realistic time-scales and programmes for project implementation;
- appropriate specifications for performance standards.

At appraisal the level of project definition is likely to be low, and therefore risk response should be characterized by a broad-brush approach. It is recommended that effort should be concentrated on:

(1) seeking solutions that avoid/reduce risk;
(2) considering whether the extent or nature of the major risks are such that the normal transfer routes may be unavailable or particularly expensive;
(3) outlining any special treatment that may need to be considered for risk transfer, for example for insurance or unconventional contractual arrangements;
(4) setting realistic contingencies and estimating tolerances consistent with the objective of preparing the best estimate of anticipated total project cost;

(5) identifying comparative differences in the riskiness of alternative project schemes.

Engineering project managers will usually have less responsibility for identifying the revenues and benefits from the project: this is usually the function of marketing or development planning departments. The involvement of project managers in the planning team is recommended, as the appraisal is essentially a multidisciplinary brainstorming exercise through which the Promoter seeks to evaluate all alternative ways of achieving these objectives.

For many projects this assessment is complex, as not all the benefits/disbenefits may be quantifiable in monetary terms. For others it may be necessary to consider the development in the context of several different scenarios (or views of the future). In all cases the predictions are concerned with the future needs of the customer or community. They must span the overall period of development and operations of the project, which is likely to range from a minimum of eight or ten years for a plant manufacturing consumer products to 30 years for a power station and much longer for public works projects. Phasing of the development should always be considered.

Even at this early stage of project definition, maintenance policy and requirements should be stated, as these will affect both design and cost. Special emphasis should be given to future maintenance during the appraisal of projects in developing areas. The cost of dismantling or decommissioning may also be significant but is frequently conveniently ignored.

2.4 Project evaluation

The process of economic evaluation and the extent of uncertainty associated with project development is illustrated by the appraisal of the hypothetical new industrial plant in Chapter 7. The use of a range of financial criteria for quantification and ranking of the alternatives is strongly recommended. These will normally include discounting techniques, but care must be taken when interpreting the results for projects of long duration.

Cost–benefit analysis

In most engineering projects factors other than money must be taken into account. If we build a dam, we might drown a historical monument, reduce the likelihood of loss of life due to flooding, increase the growth of new industry because of the reduced risk, and so on. Cost–benefit analysis provides a logical framework for evaluating alternative courses of action when a number of factors are highly conjectural in nature. If we confine ourselves to purely financial considerations, we fail to recognize our overall social objective, to produce the greatest possible benefit for a given cost.

At its heart lies the recognition that we should not ignore any factor because it is difficult or even impossible to quantify in monetary terms. Methods are available to express, for instance, the value of recreational facilities, and although it may not be possible to put a figure on the value of human life, it is surely not something we can afford to ignore.

It is essential in cost–benefit analysis to take into account all the factors that influence either the benefits or the cost of a project. Imagination must be used to assign monetary values to what at first sight might appear to be intangibles. Even factors to which no monetary value can be assigned must be taken into consideration. The analysis should be applied to projects of roughly similar size and patterns of cash flow. Those with the higher cost–benefit ratio will be preferred. The maximum net benefit ratio is marginally greater than that of the next most favoured project. The scope of the secondary benefits to be taken into account frequently depends on the viewpoint of the analyst.

It is obvious that, in comparing alternatives, we must ensure that each project is designed within itself at the minimum cost that will allow the fulfilment of objectives including the appropriate quality, level of performance and provision for safety.

Perhaps more important, the viewpoint from which each project is assessed plays a critical part in properly assessing both the benefits and cost that should be attributed to a project. For instance, if a private electricity board wishes to develop a hydroelectric power station, it will derive no benefit from the coincidental provision of additional public recreational facilities, which cannot therefore enter into its cost–benefit analysis. A public sector owner could quite properly include the recreational benefits in its cost–benefit analysis. Again, as far as the private developer is concerned, the cost of labour is equal to the market rate of remuneration, no matter what the unemployment level. For the public developer, however, in times of high unemployment the economic cost of labour may be nil, as the use of labour in this project does not preclude the use of other labour for other purposes.

2.5 Engineering risk

An essential aspect of project appraisal is the reduction of risk to a level that is acceptable to the investor. This process starts with a realistic assessment of all uncertainties associated with the data and predictions generated during appraisal. Many of the uncertainties will involve a possible range of outcome: that is, it could be better or worse than predicted. Risks arise from uncertainty and are generally interpreted as factors that have an adverse effect on the achievement of the project objectives.

Risks in engineering projects arise from a variety of sources and are of several types:

- environmental/political
- hazard/safety
- market
- technical/functional.

The implications of several of the risks likely to be encountered in engineering projects are illustrated in Figures 2.3a to 2.3d. It is relevant to note that the single-line investment curve shown in Figure 2.1 represents the 'most likely' outcome of the investment. An idea of the spectrum of uncertainty arising from the estimates and predictions is shown in Figure 2.3d. Also, the maximum risk exposure occurs at the point of maximum investment – when the project is completed and either does not function or is no longer needed.

Risks are specific to a project, are interactive, sometimes cumulative: they all affect cost and benefit.

- *Environmental risks* frequently result in compromise following comparison of cost with benefit. They are likely to have a significant influence on the conceptual design, and the response should therefore be agreed prior to sanction. Residual uncertainty may be incorporated in the analyses, usually as a contingency sum which may have to be expended.
- *Risk to health and safety* is normally considered as a hazard during design, and embraces issues such as reliability and efficiency in addition to safety. The aim is 'fitness for purpose', and the basic approach is normally compliance with industry codes. In the case of facilities that process hazardous substances a full-scale safety audit will be necessary or mandatory. This will have implications for both programme and cost.
- *Innovation.* The consequential risk of inadequate performance may be reduced by thorough testing, but appropriate time and cost provision must be included.
- *Risk to activity* relates mainly to the implementation phase of the

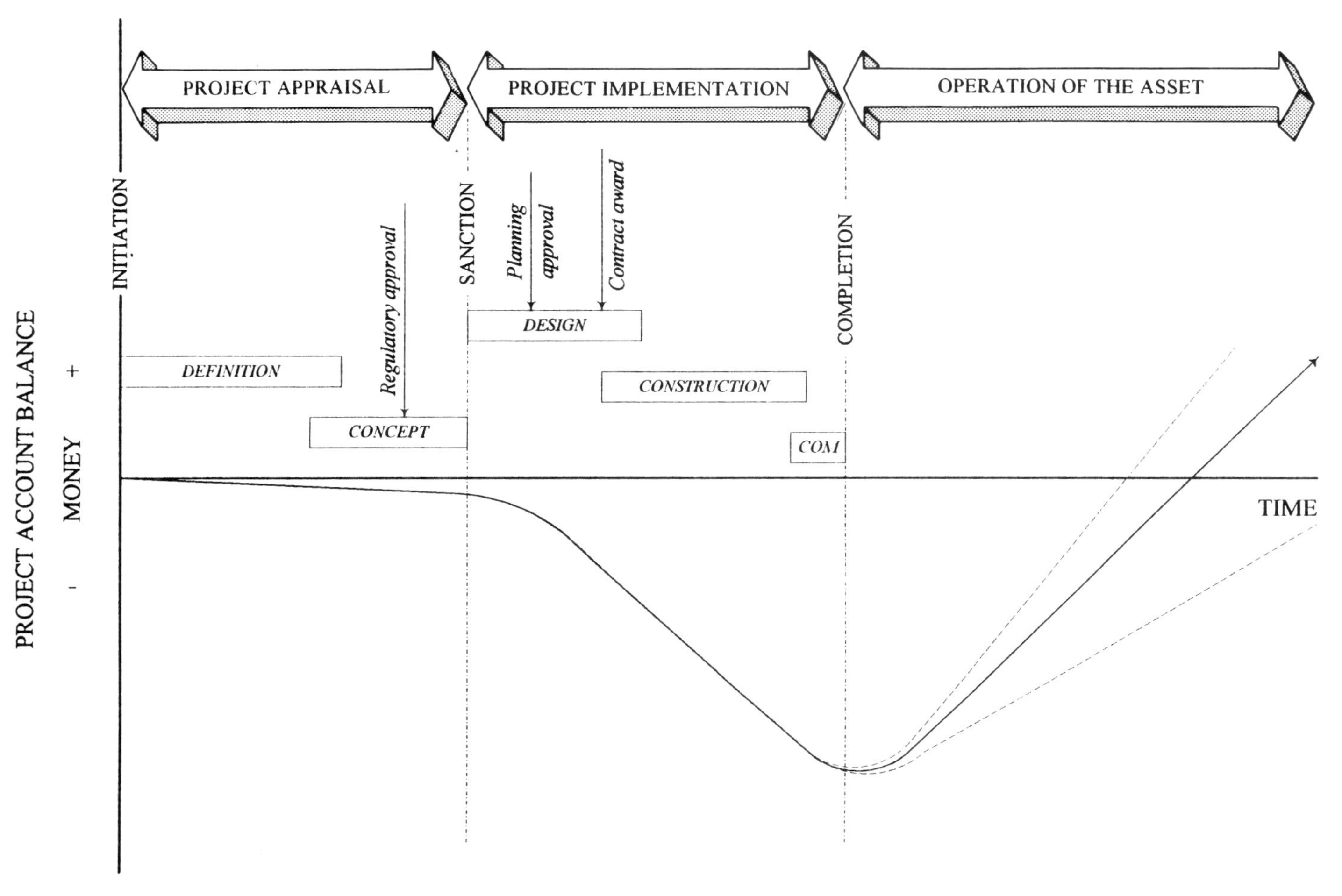

Figure 2.3(a) The effect of market uncertainty on the predicted project investment curve is shown by dotted lines.

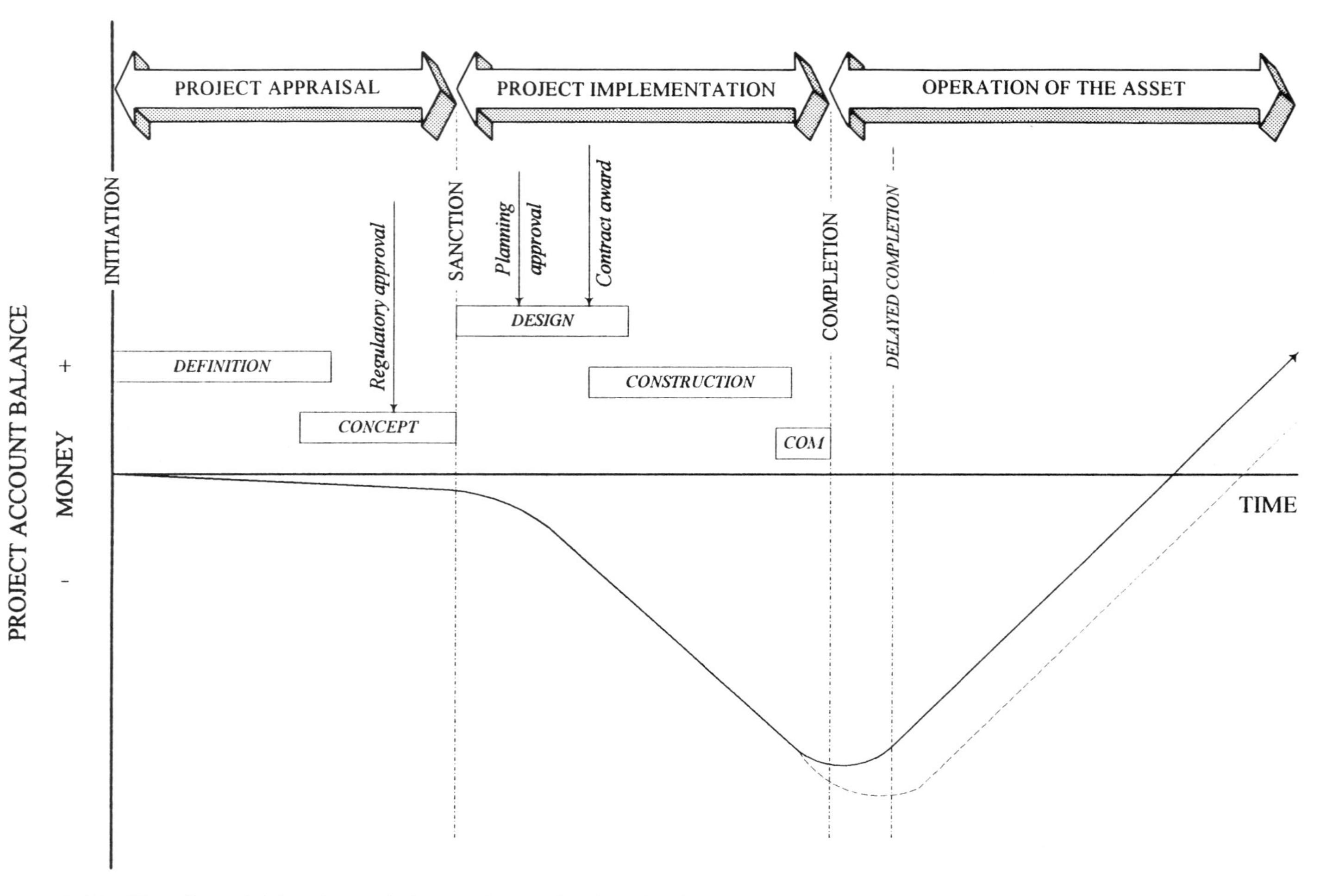

Fugyre 2.3(b) The effect of delayed completion on the predicted project investment curve.

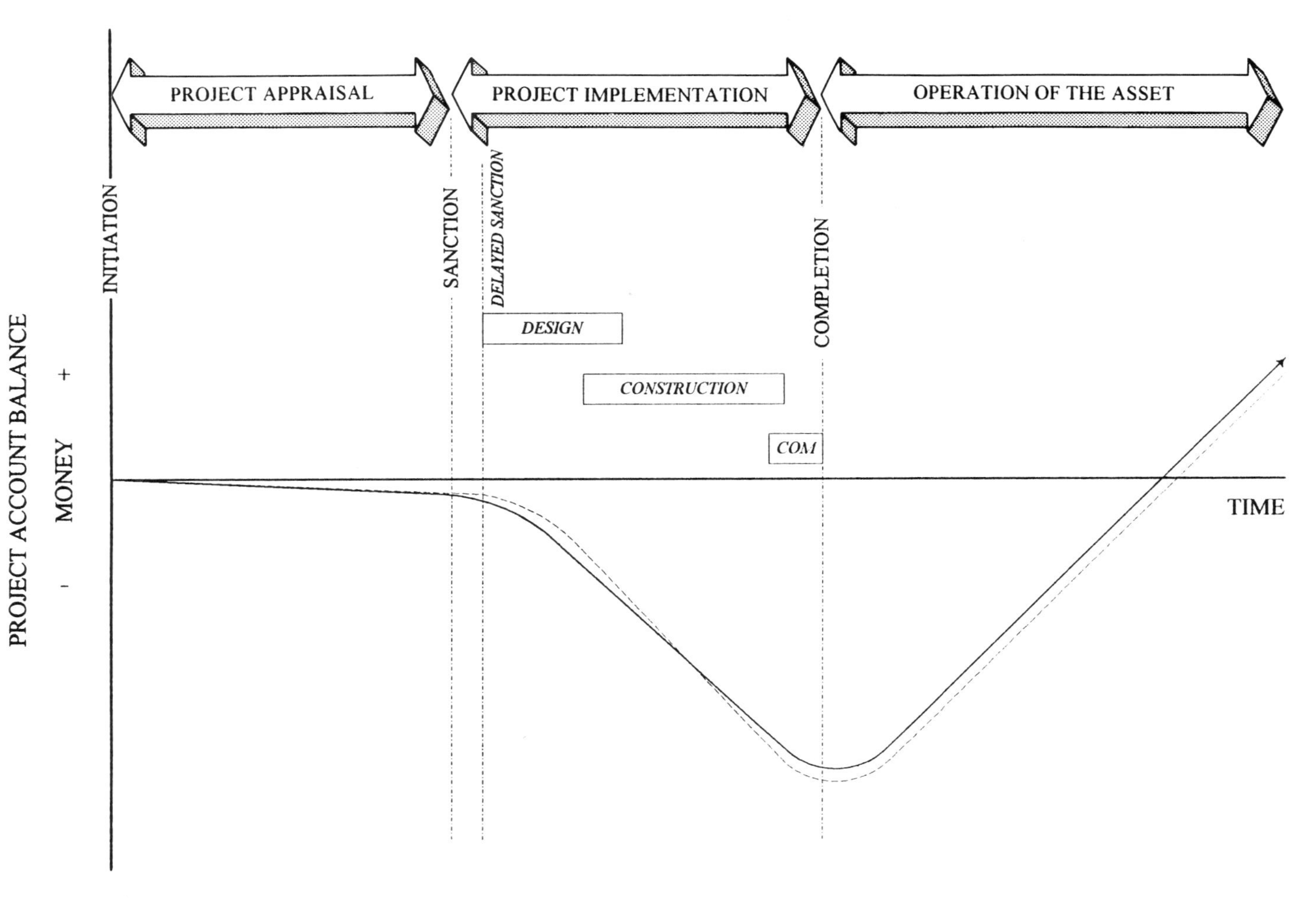

Figure 2.3(c) The possible effect of delay in sanction followed by 'fast tracking' design and construction on the predicted project investment curve.

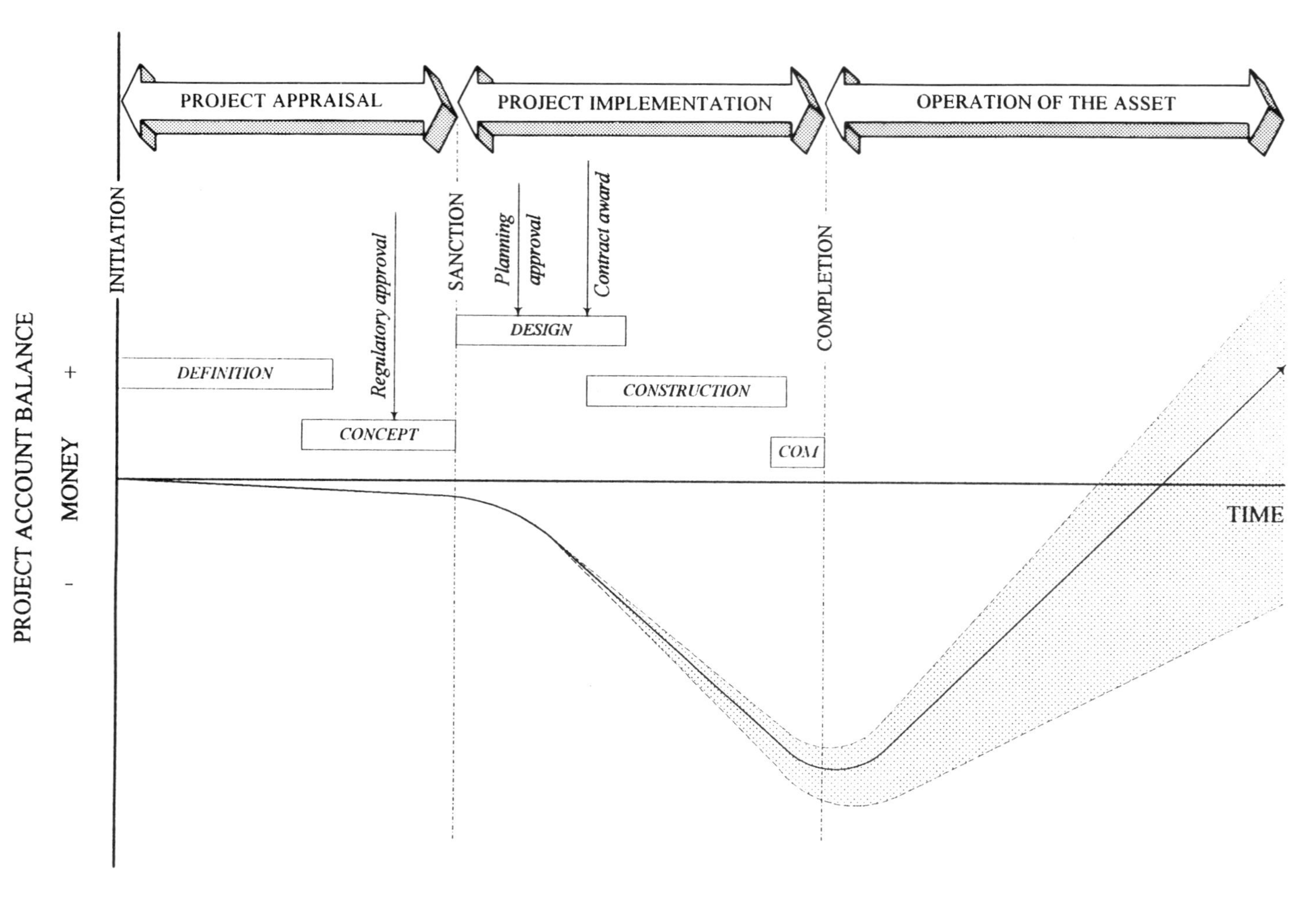

Figure 2.3(d) The cumulative effect of several variable factors is to give a spectrum of uncertainty (shaded area) about the best prediction.

project. These risks arise mainly from uncertainty and are the responsibility of the project manager, who should be allocated appropriate contingencies. The extent and nature of the contingencies depends on the magnitude and complexity of the risks and on the degree of flexibility required.

All uncertainties, particularly those that cause delay, will affect investment in the project. Many risks are associated with specific time constraints imposed on the project. The preparation of an outline programme is an essential early requirement of any approach to risk identification.

2.6 Risk management

The logical process of risk management may be defined as:

- identification of risks/uncertainties
- analysis of the implications (individual and collective)
- response to minimize risk
- allocation of appropriate contingencies.

Risk management can be considered as an essential part of the continuous and structured project planning cycle. Risk management:

- requires that you accept that uncertainty exists;
- generates a structured response to risk in terms of alternative plans, solutions and contingencies;
- is a thinking process requiring imagination and ingenuity;
- generates a realistic (and sometimes different) attitude in project staff by preparing them for risk events rather than being taken by surprise when they arise.

If uncertainty is managed realistically the process will:

- improve project planning by prompting 'what if' questions;
- generate imaginative responses;
- give greater confidence in estimates;
- encourage provision of appropriate contingencies and consideration of how they should be managed.

Risk management should impose a discipline on those contributing to the project, both internally and on customers and contractors. By predicting the consequences of a delayed decision, failure to meet a deadline, or a changed requirement, appropriate incentives/penalties can be devised. The use of range estimates will generate a flexible plan in which the allocation of resources and the use of contingencies is regulated.

Risk reduction

In general, risk may be reduced in the following ways:

- obtaining additional information
- performing additional tests/simulations
- allocating additional resources
- improving communication and managing organizational interfaces.

Market risk may frequently be reduced by staging the development of the project. All the above will incur additional cost in the early stages of project development.

Contingencies

The setting and management of contingencies is an essential part of project management. The three types of contingency are: time (float), money (allowance in budget) and performance/quality (tolerances).

Their relative magnitude will be related to the project objectives. The responsibilities/authority to use contingencies should be allocated to a named person. It is essential to know what has been used and what remains at any point in time.

The role of people

All the above risks may be aggravated by the inadequate performance of individuals and organizations contributing to the project.

Control is exercised by and through people. As the project manager will need to delegate, he/she must have confidence in the members of the project or contract team and, ideally, should be involved in their selection.

The project manager should involve staff in risk management in order to utilize their ideas and to generate motivation and commitment. The roles, constraints and procedures must be clear, concise, and understood by everyone with responsibility.

Further reading

Allen, D.M. (1980) *A Guide to Economic Evaluation of Projects*, Institution of Chemical Engineers, Rugby.

Corrie, R.I. (ed.) (1990) *Project Evaluation and Review*, Thomas Telford, London.

Misham, E.J. (1988) *Cost Benefit Analysis*, Unwin, London.

Neil, J.M. (1982) *Construction Cost Estimating for Project Control*, Prentice-Hall, New York.

Perry, J.G. and Thompson, P.A. (1992) *Engineering Construction Risks – A Guide to Project Risk Analysis and Risk Management*, Thomas Telford, London.

Chapter 3
Project Management and Quality

Quality is an important concern for all business organizations. Although the management of quality is perceived as a relatively recent concept, the Institute of Quality Assurance was established in 1919. Most project organizations now require potential partner, supplier or vendor organizations to operate a quality system. Project managers have, for many years, used procedures for certain types of project that can be adopted for the execution and integration of a quality system. This chapter describes how project management can be used effectively to develop, support and administer project-specific quality systems. To ensure safety and limit the consequences of any damage, total quality management is required.

Table 3.1 Quality assurance: ISO 9001.

- ❑ Management responsibility
- ❑ Quality system
- ❑ Contract review
- ❑ Design control
- ❑ Document control
- ❑ Purchasing
- ❑ Purchaser-supplied product
- ❑ Product identification and traceability
- ❑ Process control
- ❑ Inspection and testing
- ❑ Inspection, measuring and test equipment
- ❑ Inspection and test status
- ❑ Control of non-conforming product
- ❑ Corrective action
- ❑ Handling, storage, packaging and delivery
- ❑ Quality records
- ❑ Internal quality audits
- ❑ Training
- ❑ Servicing
- ❑ Statistical techniques

3.1 Definitions

There are many definitions of quality, usually strongly influenced by a particular industrial sector. Some are listed below.

- Quality is the ability to meet market and customer expectations, needs and requirements.
- Quality is the supplying of goods that do not come back, to customers who do.
- Quality means in conformance with user requirements.
- Quality means fitness for use.

For the purposes of this chapter it is suggested that the definition of quality, compatible with the requirements of project management, is as follows:

> 'Quality is the ability to manage a project and provide the product or service in conformance with the user requirements on time and to budget, and where possible maximizing profits.'

Currently, most business organizations now require potential partner, supplier or vendor organizations to operate a quality system. The signs are that this is increasing and those businesses that do not have a quality system will lose business opportunities. Suppliers, distributors and others providing products and services are increasingly being held responsible for any damage caused by the product to persons or property, known as product liability.

In the UK the Quality Assurance standard BS 5750 was produced in 1979 and the International Standards Organization produced ISO 9000–9004 in 1987. Table 3.1 shows the main contents of ISO 9001. There is now a compatible European equivalent, EN 29000. These documents provide guidance on the preparation and implementation of quality systems. Organizations achieving the required standards are certified for given periods of time.

3.2 Quality systems

A project quality system incorporates all stages of the project from conception to operation; sometimes until decommissioning. By focusing on the early feasibility and design stages, quality and cost requirements may be identified. In order to achieve the desired quality without unnecessary costs, a project manager must pursue an efficient system of coordinating the project activities.

The quality system should ensure that:

- the quality products and services always meet the expressed or implied requirements of the customer;
- the company management knows that quality is achieved in a systematic way;
- the customer feels confident about the quality of goods or services supplied and the method by which they are achieved.

The quality system must be adjusted to suit the project's operation and the final product. It must be designed so that the emphasis is put on preventive actions, at the same time allowing the project manager to correct any mistakes that do occur during the project life cycle.

The quality system should be based on the following activities:

- planning
- execution
- checking
- action.

All activities and tasks affecting the quality of a project require planning to achieve a satisfactory result. Execution should be based on relevant expertise and resources, and results must be checked. Checking must be followed by action. Defective products or sections of work must be removed. Information gained must be analysed and recorded in order to prevent the same defects from appearing again. Continuous upgrading of the quality system must take place. This interrelationship is shown by the Deming Circle (Figure 3.1).

Certain common rules are necessary to coordinate and direct a project's operation in order to achieve the quality objectives. These must be made known to all members of the company and project team. These rules operate in a hierarchy, illustrated in Figure 3.2, and the respective steps are described below.

- *Quality policy.* This is the main guide to the company's approach to quality matters. It clarifies the overall principles for the company's attitude towards and handling of quality management. The quality policy must not make any statement that will not have the backing and resources necessary for this achievement.
- *Quality principles.* These give more specialized guidelines as to the

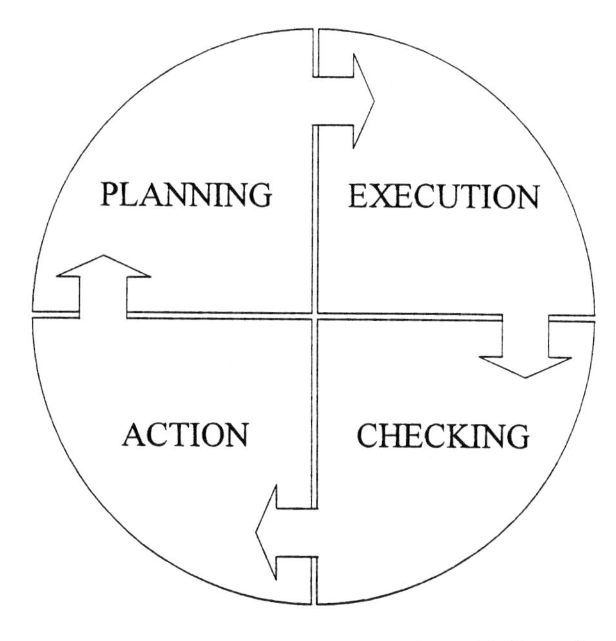

Figure 3.1 Deming Circle. (*Quality Systems – Quality a Challenge for Everyone* by Kerstin Jönson. Studentlitteratur, Lund, Sweden.)

specific project phases. In order to ensure that staff are aware of the methods of working and responsibilities, it is essential that principles and procedures are documented. This is done in the form of a quality manual.

- *Quality procedures.* These define patterns and sequences of actions to be taken, and coordinate the construction processes important to quality matters. They distribute the responsibility between different functions, those being departments within the company and members of the project team.
- *Working instructions.* These give detailed information as to the execution of the various stages of construction related to the project.

The aim of the *quality manual* is to aid the company and project team in directing their quality management. It is becoming increasingly common for customers to demand proof from suppliers of a documented quality system; the manual will also fulfil this function. It should provide an adequate description of the quality management system while serving as a permanent reference in the implementation and maintenance of that system. The quality manual is also used as a means of educating new employees by identifying the routines and responsibilities of all other employees within the company or specific to a project.

The company management and the project manager should periodi-

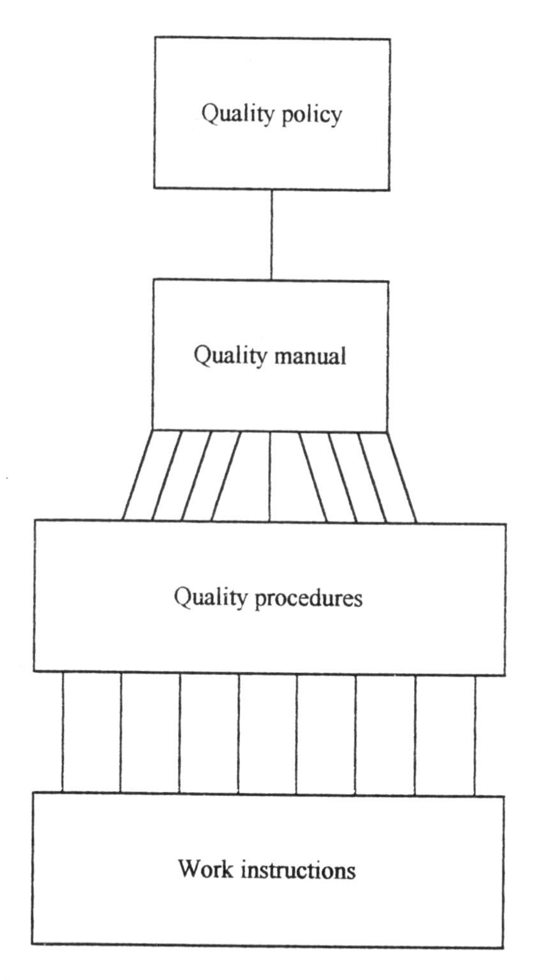

Figure 3.2 Hierarchy of quality procedures.

cally review the quality actions. A tool for this is the *quality audit*. Quality audits are similar to other in-house audits but concentrate on quality aspects. A quality audit is a systematic and independent examination to determine whether quality activities and results comply with planned arrangements, and whether these arrangements are implemented effectively and are suitable to achieve objectives. It can be directed towards goals (goal audit), systems (system audit), or structure and practicality (operation audit).

3.3 Implementation

Implementing a project-specific quality system involves the following actions.

- Determine the size of the project, the project team and its location
- Identify the status of existing documentation and determine whether or not existing documentation is compatible with the proposed project activities.
- Determine the range of project activities within the specific project.
- Identify the level of activities within the project and the method of work allocation.
- Determine the responsibilities and commitment of the project management team.

The structure of *documentation* within a quality system should take into account structure, to avoid overlapping and duplication; design, to provide a uniform simple format; distribution, to provide a manageable system that can be easily updated; and the facility for documentation changes.

A documentation system must be able to develop so that it clearly reflects the activities and methods contained in a project. Sections of the system will need to be rapidly updated and reissued to keep abreast of demands and changes in techniques and procedures.

All members of the project team should have a *job description*. The job description should define the following parameters: designation within the project, reporting procedures, key tasks and responsibilities, relationship with other members of the project team, reference to work instructions, specific performance measurement where applicable, and specified minimum training required.

It is recognized that *training* is one of the cornerstones of a quality system. Certain training is mandatory for employees and, wherever practical, should be carried out prior to any work responsibility being assigned. The following must be completed by employees involved in a project: a company introduction package and specific training regarding health and safety at work and the project quality system.

The advantages to the project manager of developing and introducing a quality system to a project are as follows.

- Profits would be increased through better management control systems.
- Quality levels would be uniform throughout the project.
- Deliveries and storage of materials would be safer.
- Costs should also be reduced through higher productivity, fewer defects and changes.
- Capital utilization can be improved through faster and safer throughput.

- The result is satisfied clients and/or promoters.

Quality costs, but lack of quality costs even more. Today's society depends heavily on reliable products and services. The consequences of failure are costly, both for society as a whole and for the project, incurring additional expense and inconvenience.

3.4 Quality-related costs

The project manager is primarily concerned with controlling time and costs, and with the inclusion of a quality system should be in a position to control time, costs and quality. The need to identify the costs of achieving quality is not new, although the practice is not widespread. In this way a monetary value can be placed on wasted effort, and on the costs of correcting these errors, and hence the work on the project can be redirected to optimize benefit.

Quality costs can be described as one of two types: operating quality costs or external assurance quality costs.

Operating quality costs

Operating costs are those costs incurred by a project in order to attain and ensure specified quality levels. These include the following.

Prevention and appraisal project costs (or investments)

- Prevention: costs of efforts to prevent failures:
  - design reviews
  - quality and reliability training
  - vendor quality planning
  - audits
  - construction and installation prevention activities
  - product qualification
  - quality engineering.
- Appraisal: costs of testing, inspection and examination to assess whether specified quality is being maintained:
  - test and inspection
  - maintenance and calibration
  - test equipment depreciation
  - line quality engineering
  - installation testing and commissioning.

Failure costs (or losses)
Internal failure costs resulting from the failure of a product or service to meet performance specification prior to delivery (e.g. product service, warranties and returns, direct costs and allowances, or product recall costs, liability costs):

- design changes
- vendor rejects
- rework
- scrap and material renovation
- warranty
- commissioning failures
- fault-finding in tests.

External assurance quality costs

External assurance quality costs are those costs relating to the demonstration and proof required as evidence by promoters, including particular and additional quality assurance provisions, procedures, data and documentation.

External quality assurance costs should be lower than operating costs, as most quality costs will have been incurred before the involvement of external quality assurance personnel. Ideally, external assurance personnel should be involved with both the project and the company.

Only by knowing where costs are incurred and their order of magnitude can project managers monitor and control them. Quality costs must be collected and recorded separately, otherwise they become absorbed and concealed in numerous overheads. Regular financial reports are vital if quality-related costs are to be visible to management.

The nature of the balance between each of these three types of quality cost and their relationship with improvement in quality means that each of the three types of quality cost reduces as quality improves, but the largest improvement is in the reduction in the cost of failure.

3.5 Quality circles

Quality circles originated in Japan. They provide a method of refining the local and project working conditions and improving productivity, based on two factors:

- the staff can often see problems that are not evident to their managers;

- the best people to fix a problem are those who stand to benefit from its solution.

There are only two ground rules that must be adhered to if a quality circle is to be successful. First, it must be implemented totally within its own resources, usually at zero cost; and second, it should tackle one problem at a time – usually the one that yields the highest benefit.

The way quality circles work is to take a small but representative team from one department. In the case of a project-specific quality circle the project manager would organize the process. Initially a brainstorming session would be held to determine any attainable improvement in a project activity. Second, the project activity would be discussed with other members of the project team. Third, a number of good ideas would be identified and proposals made on how the problem may be solved. Finally, the most inexpensive solution would be implemented within the project as a whole.

In most projects quality systems are developed as the project proceeds. Many predetermined ideas are often unsuitable to solve particular problems during the project life cycle. To ensure that problems are addressed and catered for as they occur the project manager can utilize the experience of the project team in quality circles. There can be difficulties in administering the scheme: for example, if all the key personnel on a small project are needed to form an effective quality circle it could create problems in actually organizing work on the project. Usually the problems can be overcome, but it should be recognized that a significant commitment in terms of time and resources is required if quality circles are to operate efficiently.

3.6 Quality plans

A quality plan is a document setting out the specific quality practices, resources and activities relevant to a particular process, service or project.

BS 5750: Part 02 advises that quality plans should define:

- the quality objects to be attained;
- the specific allocation of responsibility and authority during the different phases of the project;
- the specific procedures, methods and work instructions to be applied;
- suitable testing, inspection, examination and audit programmes at appropriate stages;

- ❑ a method for changes and modifications in a quality plan as projects proceed;
- ❑ other measures necessary to meet objectives.

To be of value, the first issue of a quality plan must be made before the commencement of work on site. It is also essential that it should be a document that will be read, valued and used by those in control of the work. Their preparation should be commenced at tender stage as part of the normal routine of project planning.

A typical quality plan should show the operation to be undertaken and the group with responsibility; it should identify the relevant control document and record the appropriate code numbers. Frequently, at the right-hand side of the plan there would be a column for comment and information; this is a useful element in any plan and allows additional information to be recorded and communicated effectively.

3.7 Total quality management

There are four main elements to TQM: management commitment, teamwork, techniques and the quality system (Figure 3.3). The system is how the other elements are coordinated and enabled to interact.

The main ingredients of TQM are as follows.

- ❑ It provides quality that meets the project's requirements.
- ❑ Quality is not an alternative to productivity but a means of achieving it.
- ❑ Every activity of a project contributes to the total quality of a project.
- ❑ TQM is a way of achieving project success.
- ❑ Managing quality involves systems, techniques and individuals.
- ❑ TQM is a way of managing a project.

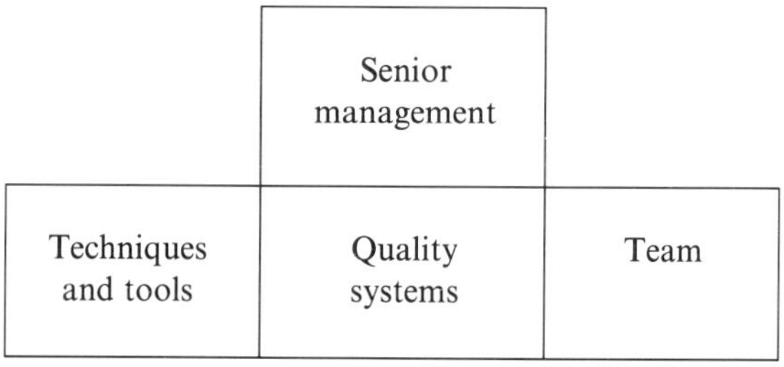

	Senior management	
Techniques and tools	Quality systems	Team

Figure 3.3 Total quality management.

The advantages of TQM in a project are as follows.

- Quality in meeting the project specification saves money.
- TQM alleviates the quality costs that result from poor quality.
- Costs are reduced by preventing poor quality.
- Capacity is increased by improving quality.
- The system provides a basis for teamwork and techniques to interact.
- The system recognizes the need to balance risk, benefit and cost.
- The system allows for changes in the project.
- The system records all activities in the project.

Many projects are unique and require the development of quality systems specific to a project. In many cases existing quality systems can be used as the basis of a project-specific system with the addition of control documentation covering activities not previously encountered. Many project-specific systems are developed as a project proceeds, requiring continuous updating of the system.

A quality system is itself a management system, which can be used by a project manager to ensure that the project meets the quality required through an efficient management control system.

Further reading

Ashford, J.L. (1989) *The Management of Quality in Construction*, E & FN Spon, London.

BS 5750: 1987, *Quality Systems*, British Standards Institution, London.

BS 6143 (1992) *Guide to the Economics of Quality*, British Standards Institution, London.

ISO 9000: 1987, *Quality Systems*.

Jönson, K. *Quality Systems – Quality a Challenge for Everyone*, Studentlitteratur, Lund, Sweden.

Oakland, J.S. (1989) *Total Quality Management*, Heinemann Professional Publishing, Oxford.

Chapter 4

Project Appraisal and Environmental Impact

In the past the main tools for project appraisal were cost–benefit analysis and cost-effectiveness analysis: both logical and quantitative methods for identifying whether or not a project was worth implementing. Many publicly funded projects, such as water supply systems, were assumed to have such overwhelming benefits that only the costs were assessed in order to determine which of the various alternative methods of achieving the project objectives was the most cost-effective. It is now recognized that all projects will result in some unquantifiable costs and benefits, and therefore today these appraisal methods are supplemented by environmental impact analysis and assessment. By its nature, environmental impact assessment (EIA) requires a much more qualitative approach to project appraisal than cost–benefit analysis, although this is gradually changing as new methods for valuing environmental impacts emerge and develop.

This chapter covers the main elements of EIA as it is currently used for the appraisal of projects in many countries. New developments in the evaluation and 'monetizing' of environmental impacts are reviewed and discussed in the latter part of the chapter.

4.1 Environmental impact

Before discussing environmental impact it is first necessary to define what is meant by environment, as it can mean a variety of things to different people – from the light and heat inside a building, to the outdoor environment of natural woodlands, moors, rivers and seas, to the ozone layer surrounding the planet. In the context of the environmental impact of projects, and the discipline of environmental impact assessment, the term 'environment' is taken in its widest context to include all the physical, chemical, biological and socioeconomic factors that influence individuals or communities. It not only includes the air, water and land, and all living species of plants, animals, birds, insects, and micro-organisms, but also man-made artefacts and structures, and factors of importance to the social, cultural, and economic aspects of human existence.

In this context, all projects have an effect or impact on the environment – indeed it could be argued that unless they did there would be no point in implementing projects at all. Some of these may be seen as positive impacts (benefits), while others will have a detrimental effect (costs). The purpose of environmental impact assessment is to evaluate these positive and negative effects as objectively as possible and present the information in a manner that is accessible to decision makers so that it can be used in conjunction with other appraisal tools, such as cost–benefit analysis.

Engineering projects may have an impact on the full range of environmental features including air, water, land, ecology, sound, human aspects, economics, and natural resources. Many of these impacts can be measured in terms of changes to specific quality parameters such as concentration of particulates or hydrocarbons in the air, or dissolved oxygen concentration in water. Other impacts, such as the aesthetic qualities of a landscape or a structure, or the importance of preserving an historic building, are not so easily quantified.

Some of the impacts will be directly attributable to the project – such as noise from an airport or road, the visual impact of a structure, or the pollution of the air or water from a factory. These are referred to as *direct* or *primary impacts*. Other impacts may arise indirectly through the use of materials and resources required for the project. Examples include the pollution of the air from the manufacture of cement used for a project, or the impact of quarrying for raw materials needed for the project. These impacts may affect areas remote from the project itself and are termed the *indirect* or *secondary impacts*. Lower-order impacts can also be identified, such as the impacts on the environment caused by the manufacture of equipment used for quarrying or cement manufacture.

The range of environmental impacts arising from a project will continue throughout the operating life of the project, and can therefore be regarded as permanent or long-term impacts. The pollution of the air caused by the operation of a thermal power station is thus an example of a long-term impact. Short-term or temporary impacts, on the other hand, are those arising from the planning, design, and construction phases of the project. Typically these might include the noise and dust generated by the construction process itself, or temporary changes to water table levels, for example.

Environmental impacts may therefore be temporary and direct, temporary and secondary, permanent and direct, or permanent and secondary, and, as discussed later in this chapter, it is necessary to differentiate between these categories of impact within an environmental impact statement.

4.2 Environmental impact assessment

Environmental impact assessment (EIA) is a logical method of examining the actions of people, and the effects of projects and policies on the environment, in order to help ensure the long-term viability of the earth as an habitable planet. EIA aims to identify and classify project impacts, and predict their effects on the natural environment and on human health and well-being. EIA also seeks to analyse and interpret this information and communicate it succinctly and clearly in the form of an *environmental impact statement* (EIS), which can be used by decision makers in the appraisal of projects.

The natural environment is not in a completely steady state; changes occur naturally over time, some extremely slowly but others at a much faster rate. Therefore any study of the impacts of a project on the environment must be seen in the context of what would have happened if the project had not been implemented. The environmental impact is thus any environmental change that occurs over a specified period, and within a defined area, resulting from a particular action, compared with the situation as it would have been if the action had not been taken.

The report of all environmental impacts that are predicted to arise from a particular project is normally termed an EIS, and is now often required in order to obtain sanction for a project. In many countries this procedure has now been incorporated in the legal processes of obtaining project and planning approval, and in other situations it has been used voluntarily as part of general project preparation and evaluation.

EIA is a process rather than an activity occurring at one point in time during the project cycle. The various steps in the process are shown in Figure 4.1. The process starts with *screening*, which should be done in the early stages of a project, as it is concerned with determining whether or not a detailed EIA is required or necessary.

Scoping is the second step of the EIA process and is essentially a priority-setting activity. It may be seen as a more specific form of screening, aimed at establishing the main features and scope of the subsequent environmental studies and analysis. The results of the scoping exercise provide the basis and guidelines for the next step in the process – the baseline study and impact assessment itself.

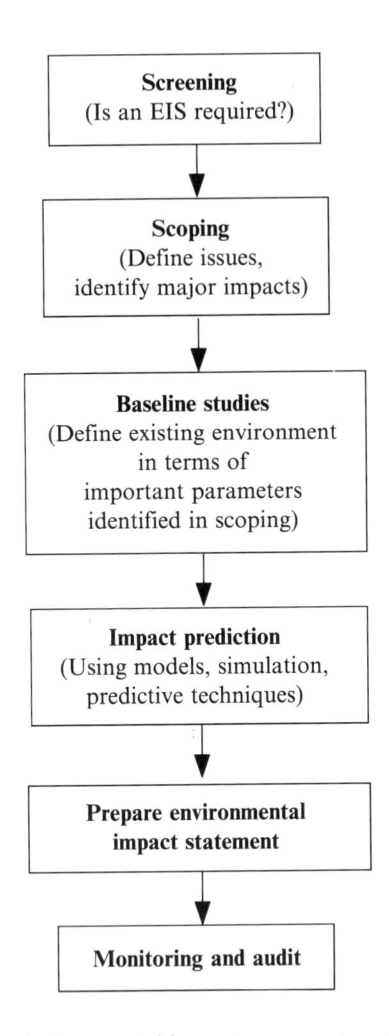

Figure 4.1 Environmental impact assessment process.

Scoping sets the requirements for the *baseline study*, which is the collection of the background information of the ecosystem and the socio-economic setting of a proposed development project. The activities involve the collection of existing information and the acquisition of new data through field examination and the collection of samples. This step can be the most expensive step of the EIA because it needs a large number of expert people to carry out field surveys and analysis.

The baseline study and the project proposals are then used as the basis for predicting how environmental parameters will change during both the construction and operational phases of the project. Environmental and social scientists will be required for this step, as often sophisticated modelling and simulation techniques must be used.

The predictions and the baseline study are then used to prepare the report – the EIS – which is used in the appraisal and approval processes. Further details of the structure of the report and the more common methods of presenting EIA information are given in subsequent sections of this chapter.

Because of the uncertainty associated with environmental impact predictions, it is important that the major environmental parameters are monitored throughout the implementation of the project. This can provide valuable information for future EIAs and generally improve the accuracy of forecasting models and methods. The process of comparing the impacts predicted in an EIA with those that actually occur after the implementation of the project is referred to as *auditing*.

4.3 Screening

Screening is the first step of the EIA process; it is the selection of those projects that require an EIA. For example, construction of a simple building rarely requires an EIA, whereas the construction of a large coal-burning power station would need a full-scale EIA to be done and an EIS to be prepared, which would include the results of the required scientific and technical studies, supporting documentation, calculations, and background information.

The extent to which the EIA is needed for a project is also defined in this step, as the impact of a project depends not only on its own specific characteristics but also on the nature of the environment in which it is set. Therefore the same type of project can have different impacts, or a different intensity of some impacts, in different settings. So, for any particular project, a full-scale EIA may be required for one site but not for another.

The requirements for an EIA also largely depend on the legislative policy of the country in which the project is to be carried out. Through the screening process, only those projects that require EIA are selected for study. By doing this, the unnecessary expenditure and delay is avoided.

The screening process will depend on whether or not appropriate legislation exists, and where it does the laws will often differentiate between those types of project that need an EIA and those that do not. The legislation may also define the extent of the study that will be needed for a particular type of project. Where there is no appropriate law relating to EIA, or if the types of project requiring EIA are not specifically defined, the project manager will have to undertake a preliminary study to ascertain the most significant impacts of the project on the environment. On the basis of the results of this study and, after studying the environmental law and assessing the political situation of the country, the project manager will have to decide whether a full-scale EIA is needed.

4.4 Environmental legislation

The concept, and eventually the practice, of EIA evolved, developed, and was incorporated into legislation in the USA in the 1970s. The idea was subsequently introduced into other countries, such as Canada, Australia, and Japan, but it was not until 1985 that EIA was fully accepted within the European Community. At that time a European Community Council Directive 'on the assessment of the effects of certain public and private projects on the environment' was issued. This required all Member States to implement the recommendations set out in that Directive by July 1988.

The Directive called for governments of member states to enact legislation that would make EIA mandatory for certain categories of project, and recommended for other types of project. The projects where an EIA is mandatory include:

- crude oil refineries;
- gasification and liquefaction of coal and shales;
- thermal power stations;
- radioactive waste storage and disposal facilities;
- integrated steel and cast-iron melting works;
- asbestos extraction and processing works;
- production of asbestos products;
- integrated chemical installations;
- motorways and express roads;
- long-distance railways;
- airports with runways greater than 2100 m;
- trading ports and inland waterways and ports for traffic over 1350 tonnes;
- waste disposal for the incineration, chemical treatment or land fill of toxic and dangerous wastes.

The categories of project where an EIA is recommended include the following:

- *Agriculture.* For example, poultry and pig rearing, salmon breeding, land reclamation, water management, afforestation, restructuring of rural land holdings, use of uncultivated land for intensive agricultural purposes.
- *Extractive industries.* Deep drilling, extraction of sand, gravel, shale, salt, phosphates, potash, coal, petroleum, natural gas, ores, and installations for the manufacture of cement.
- *Energy industries.* Production of electricity, steam and hot water; installations for carrying gas, steam, and hot water; overhead cables for electricity; storage of fossil fuels; production and processing of nuclear fuels and radioactive waste; hydroelectric installations.
- *Processing of metals.* Iron and steel works, non-ferrous and precious metals installations, surface treatment and coating of metals, manufacture and assembly of motor vehicles, shipyards, aircraft manufacture and repair; manufacture of railway equipment.
- *Manufacture of glass.*
- *Chemical industries.* Production of pesticides, pharmaceuticals, paint, varnishes, peroxides, and elastomers; storage of chemical and petrochemical products.
- *Food industries.* Manufacture of oils and fats, packing and canning, dairy product manufacture, brewing, confectionery, slaughterhouses, starch manufacturing, sugar factories, fish-meal and fish oil factories.
- *Textile, leather, wood, and paper industries.*
- *Rubber industries.*
- *Infrastructure projects.* Urban and industrial estate development, cable cars, roads, harbours, airports, flood relief works, dams, water storage facilities, tramways, underground railways, oil and gas pipelines, aqueducts, marinas.
- Hotels, holiday villages, race and test tracks for cars and motor cycles, waste disposal installations, wastewater treatment plants, sludge disposal, scrap iron storage, manufacture of explosives, engine testbeds, manufacture of artificial fibres, knackers' yards.

In the UK, these requirements for EIA have been incorporated into the existing planning legislation, and the developer or initiator of a project now has to supply the relevant environmental impact information in order to obtain sanction for the project. Prior to the enactment of the new legislation, planning procedures only addressed land use considerations, but now issues of pollution control and environmental impact must also be considered.

Environmental impact issues must also be addressed in many overseas projects, either as a result of relevant legislation having been introduced by particular countries or, perhaps more importantly, because bilateral funding agencies (such as the UK's Overseas Development Administration) and multilateral funding agencies (such as the World Bank and the Asian Development Bank) now have requirements for EIAs built into their loan and grant approval procedures.

4.5 Scoping

The US National Environmental Protection Agency Council on Environmental Quality defined scoping as 'an early and open process for determining the scope of issues to be addressed and for identifying the significant issues related to a proposed action'. Scoping is therefore intended to identify the type of data to be collected, the methods and techniques to be used, and the way in which the results of the EIA should be presented.

The question of what is a significant impact is not always an easy one to answer, and requires judgement, tact, and an understanding of not only the technical issues, but also the social, cultural, and economic ones. It is likely to involve value judgements based on social criteria (aesthetic, human health and safety, recreation, effect on lifestyles), economic criteria (the value of resources, the effect on employment, the effect on commerce), or ethical and moral criteria (the effect on other humans, the effect on other forms of life, the effect on future generations).

The baseline study and the impact assessment are expensive and time-consuming, and the determination of the impacts of a proposed project or policy depends on the current state of knowledge relating to any particular aspect of the environment. Very often this depends on the latest developments in the science and technology of environmental monitoring and prediction as well as the professional experience and judgement of those who carry out the EIA. It is therefore important to ascertain what can be reasonably carried out within the allowed time and budget in order that best use can be made of available resources. This is one of the tasks of the scoping exercise. The other main task is that of determining which are the most important impacts that may occur, and this will vary with the environment surrounding the project. As it is the people living in or near the area of a proposed project who are most likely to have detailed local knowledge and to be most concerned about the local environment, it is extremely important to solicit local public opinion about the project and its perceived impacts on the environment. This is best done at the scoping stage, as local knowledge is a valuable source of data for the baseline studies.

There are two major steps to take in seeking public opinion, the first of which is to identify the affected population/target group and attempt to obtain their opinions. Local political and environmental concern groups should be consulted at an early stage to avoid unnecessary confrontation at a later stage. In order to carry out this public consultation exercise a schedule of meetings, with the affected population and target groups, needs to be arranged. Care should be taken to ensure that all affected and interested groups are included in the programme. At these consultation meetings, the objectives, possible impacts and activities associated with the project should be explained, and everyone should be encouraged to express their opinions on the project and its possible impacts, and to suggest mitigation measures. Sufficient time should be allowed for these meetings, and all suggestions and comments should be recorded in meeting minutes and made public. Major impacts and areas of concern may be identified from this public consultation exercise.

This opinion and data collection procedure is quite a time- and money-consuming activity, and requires patience and diplomacy. In many cases, the EIA has to be completed in a very short period of time, so that meetings with the affected people and other interested groups can become very difficult, if not impossible. In such cases, sample questionnaire surveys can be carried out and, by using statistical methods, probable results may be formulated. Where appropriate, it may be possible to carry out some of these surveys by telephone.

A method used in Canada to ensure the public's involvement is to use a panel of four to six experts, who are selected to examine the environmental and related implications of a particular project. The panel is then responsible for issuing guidelines for preparing the EIS and reviewing it after completion. The panel will collect public views through written comment, workshops or public meetings before completing the guidelines for the EIS.

The advantage of scoping is that it helps to obtain advance agreement on the important issues to be considered in the following activities of the EIA and helps to use scarce resources efficiently to analyse the impacts by preventing unnecessary investigation into the issues/impacts that have little importance. Like good early project planning, scoping can save enormous expense and time during the later and most costly steps of EIA. Good scoping requires planning, using competent staff, and providing adequate resources.

The outcome of a scoping exercise should be a list of priority concerns and guidelines for the preparation of the EIS, including a structure for the baseline study.

4.6 Baseline study

The scoping exercise should provide information about which data are required and which data are unnecessary. If this information is not provided by the scoping team, a great deal of time and money can be wasted in gathering irrelevant and unnecessary data. Unfortunately there are a large number of examples of EIA baseline studies that have included a lot of fairly irrelevant material simply because those preparing them have included all the data that were available, even if they were unnecessary. The baseline study should concentrate on gathering the important data as identified in the scoping exercise. Some of this will be available from previous studies or other sources, but it may be necessary to conduct extensive monitoring exercises to gather other needed data. The thing to avoid is costly gathering of data that are of little or no importance.

The baseline study should result in a description of the existing environment, and it should be remembered that this is not static; changes will occur even without the proposed project. Information should be relevant to the impacts discussed, and one should be selective in the information used to describe the baseline. For example, there is no need to spend a great deal of time and energy on data collection and description of present and future noise levels, if the project is very unlikely to generate any additional noise. The baseline description should be confined to those impacts and parameters that matter. Too often the focus of the baseline survey has been on what is available rather than on what is needed.

The outcome of the baseline study is a form of environmental inventory based on primary environmental data collection where appropriate, the opinions of preselected individuals and groups, and other sources, such as the monitoring and evaluation data from similar completed projects.

As the impacts of the proposed project on the environment are forecast on the basis of the data collected during the baseline study, it is important that the study is carried out accurately and thoroughly.

4.7 Impact prediction

For each alternative, the consequences or impacts should be predicted using the most appropriate methods, which may include predictive equations, modelling and simulation techniques. The predictions should include, and differentiate between, primary, secondary, short- and long-term effects. There should be an attempt to show the effect of all project activities on a comprehensive range of environmental parameters, which should have been identified during the scoping stage. These parameters may be grouped into various categories such as biological, physical, social, economic, and cultural, and wherever possible quantitative predictions should be made.

Decision makers need to address the question of the accuracy of these predictions – imprecise predictions can generally be counted on to be fairly reliable, whereas there is often a degree of uncertainty associated with precise forecasts. For example, if a power station is being considered, an imprecise prediction such as 'the levels of particulate concentration in the air in the immediate vicinity will increase' is likely to be very reliable, whereas making a precise prediction of the value of the particulate concentration will almost certainly suffer from inaccuracies. Some assessment of the accuracy of predictions could be included.

4.8 Environmental impact statement

The EIS itself is a detailed written report on the impact of the project on the environment and should contain the following:

- The need for the project: a description of the aims and objectives of the project, explaining clearly why the project is required.
- The baseline study report: a comprehensive description of the existing environment, indicating the levels of important environmental parameters, quantitatively wherever possible.
- A list and description of all reasonable and possible project alternatives, including the 'do nothing' option. Even options that might be considered 'non-viable' should be mentioned, and reasons given for their rejection. For each of the main alternatives, the following should be included:
  - a clear description of the project during construction and operation, giving details of the use of land, materials, and energy, and estimates of expected levels of pollution and emissions;
  - a clear estimate of the environmental consequences of each alternative, including predictions of the effects of all the significant impacts on human health and welfare, water, land, air, flora, fauna, climate, landscape, and cultural heritage. This section provides the scientific and technical evidence on which comparisons between the alternatives can be made. The report should clearly draw the reader's attention to any serious adverse

  impacts that cannot be avoided or mitigated, as well as any irreversible environmental consequences and irretrievable use of natural resources;
  - the severity of each impact and a clear description of the methods used for measuring and predicting it.
- The environmental consequences of each alternative should be compared, possibly in tabular or other easily readable form, as discussed later in this chapter. Many consider this to be the most important section of an EIS, and it may often be the first section to be referred to, as it provides an easily understood comparison of all alternatives and their main environmental consequences.
- The statement conclusions should indicate the preferred option and describe any mitigation measures that may be required to minimize environmental impacts of this option.
- A non-technical summary should be included.
- The report should include an index and any necessary appendices for detailed calculations and data analysis.

4.9 Presenting EIA information

One of the difficult tasks in preparing an EIS is that of presenting the information in a way that is comprehensive, yet readily understood by decision makers. This is particularly true of comparisons of the severity of impacts of alternative proposals. A number of commonly accepted ways of doing this have emerged, and because of their critical importance within the EIS, they are often referred to as EIA 'methods'. These include a number of different forms of check-lists, overlay mapping techniques, networks, multi-attribute utility theory, and matrix methods.

Check-lists

Check-lists are one of the oldest forms of EIA method and are still in use in many different forms. Usually check-lists consist of lists (prepared by experts) of environmental features that may be affected by the project activities.

Sometimes a list of project actions that may cause impacts is also incorporated. Check-lists may be a simple list of items or more complex variations that incorporate the weighting of impacts. There are five types of check-list commonly in use:

- *Simple lists.* A list of environmental factors and development actions, with no guidance on the assessment of the impacts of the project on these factors. The principal use of this method is to focus attention and to ensure that a particular factor is not omitted in the EIS.

- *Descriptive check-lists.* This type of check-list gives some guidance on the assessment of impacts, although it does not attempt to determine the relative importance of impacts.
- *Scaling check-lists.* These consist of a list of environmental elements or resources, accompanied by criteria for expressing the relative value of these resources. For each alternative being assessed, the appropriate criteria are selected, and any impact that exceeds the defined limit is considered to be significant and should be highlighted for the decision makers' attention.
- *Scaling–weighting check-lists.* A panel of experts decides on the weight to be assigned to each environmental parameter. The idea is to assign a weighted score or scale to each possible impact to enable one impact to be compared with another, and to provide an overall environmental impact score for each alternative.
- *Questionnaire check-lists.* This type of check-list consists of a series of direct linked questions which are posed to a variety of professionals, who are asked to respond yes, no, or unknown to each of the questions.

Overlay mapping

This procedure consists of producing a set of transparent maps showing environmental characteristics (physical, social, ecological, or aesthetic) of the proposed project area. A composite characterization of the regional environment is produced by overlaying these maps, and different intensities of the impacts are indicated on the maps by different intensities of shading.

This method is suitable for the selection of routes for new highways or electrical transmission lines, for example.

One of the main difficulties in using this method is that of superimposing all the maps in a comprehensive way, especially when the number of maps is large. However, by using computers, these problems can be overcome, and a number of computer packages have been developed for this purpose.

Networks and systems diagrams

These methods were developed for the identification of secondary, tertiary and higher-order impacts from an initial impact. The methodology starts from a list of project activities to establish cause–condition–effect relationships – in the case of systems diagrams these effects are quantified in terms of energy flows. It attempts to recognize a series of impacts that may be triggered by a project. By defining a set of possible networks, the user can identify impacts by selecting the appropriate project actions.

The advantage of these approaches is that they can show the interdependence of parameters, and the effects of changes in one parameter on other parameters. The limitations are that they are only really suitable for the assessment of ecological impacts, and they are expensive, time-consuming, and require periodic updating. Systems diagrams depend on a knowledge of the ecological relationships in terms of energy flow, which is often very difficult to characterize.

Multi-attribute utility theory

Utility theory in EIA has been applied most often to site selection, especially for projects such as major power stations, and waste disposal facilities. Alternate projects have different environmental impacts and exhibit different levels (intensity) of the same impact. These methods provide a logical basis for comparing the impacts of alternatives to aid in decision making. The methodology follows the steps indicated below.

(1) Determine the environmental parameters that may be affected and which can be measured. For each parameter, use appropriate prediction methods to estimate the value of each parameter after the proposed project has been implemented.

(2) Establish the desirability or otherwise of different levels of each parameter, and formulate the utility function through systematic comparison of those different levels. Utility $U_i x_i$, of each parameter x_i, is measured on a scale of 0 to 1, where 0 is the lowest utility (i.e. the worst possible situation) and 1 is the highest utility (the best possible situation).

    This step is highly subjective, and relies on the values, knowledge, and experience of 'experts'.

(3) Determine a scaling value for each of the environmental parameters which will reflect the relative importance as perceived by decision makers. This is denoted by k_i.

    Again, the scaling factors are highly subjective and value-laden.

(4) The above steps have to be carried out for each of the alternatives separately, and the total utility, or a composite *environmental quality index* (EQI) for each of the alternatives, is then calculated as shown below.

$$\text{EQI} = U(x) = \sum_{i=1}^{n} k_i \, U_i(x_i) \qquad \textbf{(4.1)}$$

where k_i = scaling factor of parameter x_i
$U_i(x_i)$ = utility function of parameter x_i
$U(x)$ = multi-attribute function.

(5) Determine which alternative which has scored the highest EQI value. This will be the least environmentally damaging and the best of the alternatives.

One of the advantages of this method is that the concepts of probability and sensitivity analysis can be incorporated into it and, using computers, a number of 'what if' scenarios can be examined. The fact that a single number – the EQI – is produced is seen by some as an advantage, because it is easy for the decision maker to compare one alternative with another. Others see this as a major drawback, because it produces what appears to be an objective numerical comparison, but which is in fact based on hidden subjective assessments of the utility factors and the scaling factors. Another criticism of the method is that it militates against public understanding and involvement by being unnecessarily technocratic and complex.

Finally, a further limitation of this methodology is that it assumes that the environmental parameters are independent of each other, which is not the case, and that they are fully dependent on the probability assumptions.

Matrices

Simple interaction matrix

This form of matrix is simply a two-dimensional chart showing a checklist of project activities on one axis and a checklist of environmental parameters on the other axis. Those activities of the project that are judged by experts to have a probable impact on any component of the environment, are identified by placing an X in the corresponding intersecting cell. Matrix methods were originally proposed and developed by Leopold and his colleagues of the US Geological Survey to identify the impacts on the environment for almost any type of construction project. Their matrix, now known as the Leopold matrix, consists of 100 specified actions of the project along the horizontal axis and 88 environmental parameters along the vertical axis.

A simple two-dimensional matrix is, in effect, a combination of two check-lists. It incorporates a check-list of project activities on one axis and a check-list of environmental characteristics, which may be affected by the actions of the project, on the other axis. From the matrix, the cause–effect relationships of actions and impacts can be identified easily by marking the relevant cells. Various modifications have been made to Leopold's original lists of activities and environmental characteristics to suit particular situations.

Quantified and graded matrix

The original Leopold matrix is slightly more complex in that a number, ranging from 1 to 10, is used to express the magnitude and importance of the impacts in each cell. Thus a grading system is used in place of the simple X.

Modification of matrices has been going on over time with the advancement of EIA. In 1980, Lohani and Thanh suggested another grading system to incorporate the relative weights of each development activity. This method can identify major activities and areas that need more attention.

The advantages of the matrix are that it can rapidly identify the cause–effect relationships between impacts and project activities and can express the magnitude and importance in both qualitative and quantitative forms. Another major advantage of the matrix method is that it can communicate the results to the decision makers as a summary of the EIA process. It can be used for any type of project and can be modified as required.

4.10 Monitoring and auditing of environmental impacts

Monitoring

EIA is largely concerned with making predictions about the effects that projects will have on the environment. Unless the actual impacts are measured and monitored during and after project implementation, and compared with those predicted, it is not possible to assess the accuracy of predictions or to develop improved methods of prediction. The monitoring of environmental impacts is thus an essential component of the EIA process.

The aim of monitoring is first to detect whether an impact has occurred or not, and if so to determine its magnitude or severity, and second to establish whether or not it is actually the result of the project and not caused by some other factor or natural cause. In order to do this it is desirable to identify 'control' or 'reference' sites, which should be monitored as well as the project site. The reference site should be as similar as possible to the project site, except that the predicted project impacts are unlikely to occur. To be effective, monitoring should begin during the 'baseline' study and continue throughout the construction and operational phases of the project.

During the scoping stage, a framework for monitoring should be established. This should outline which parameters to monitor, the required frequency of sampling, the magnitude of change of each parameter that is statistically significant, and the probability levels of changes occurring naturally. The establishment of this framework is perhaps more important than the actual details relating to data measurement and collection, as there is little point in spending a great deal of time and money in gathering irrelevant or insignificant data. The monitoring process can be very expensive.

While the main objective of monitoring is to assess the validity and accuracy of predictions (auditing), it is worth noting that monitoring data can provide an early warning of possibly harmful effects in time to initiate mitigating measures and thus minimize adverse impacts. Monitoring improves general knowledge about the environmental effects of different types of project on a variety of environment parameters. As more monitoring exercises are carried out, the data available will increase and become more reliable, thus improving future EIA studies.

Auditing

Auditing is the process of comparing the predicted impacts with those that actually occur during and after project implementation. Auditing thus requires accurate and well-planned monitoring, as explained above.

Auditing and monitoring are steps in the EIA process that are often disregarded or not given sufficient attention. One of the reasons for this is that EIA is seen by many as a hurdle in the approval and sanction phase of a project and, once this has been granted, the environmental effects are given much less attention. If EIA is to be used as a method of reducing environmental impacts and/or mitigating them, the auditing process must be given a much higher priority than is currently the case. Auditing not only helps to provide information on the accuracy of environmental impact predictions, but also highlights best practice for the preparation of future EISs.

Audit studies of EIAs produced in the past have shown considerable variation and inaccuracies in environmental parameter predictions, and a large number of impacts studied could not be audited for a variety of reasons. These included inappropriate forms of prediction, the fact that design changes had been made after the EIS had been completed, making some predictions invalid, and inadequate or non-existent monitoring. In some cases less than 10% of the impacts evaluated in an EIA could be audited, and of these less than 50% were shown to be accurate. There is therefore room for improvement in the preparation and presentation of EISs, and effective monitoring and auditing are absolutely essential if these improvements are to be made.

4.11 Environmental economics

Decision makers are often faced with two documents: the mainly qualitative report on environmental effects (the EIS), and the 'objective' numerical statement of costs and benefits (the cost–benefit appraisal). Both are concerned with costs and benefits, but they are written and presented in different terms and forms. For this reason, a relatively new science of *environmental economics* has developed, which helps to coordinate these two appraisal tools and perhaps eventually to combine them.

Cost–benefit analysis (CBA)

In its simplest form, the principles of CBA can be expressed by the following equation:

$$\Sigma(B_t - C_t)(1 + r)^{-t} > 0 \qquad (4.2)$$

where B_t = benefits accruing at time t
C_t = costs incurred at time t
r = discount rate
t = time

As mentioned previously, only those costs and benefits that can be easily quantified are normally included, and the environmental and social costs and benefits are ignored. There are two questions to address, the first being whether or not there are indeed any real costs and benefits that can be ascribed to the environment, and the second being how can they be valued in the same terms as other project costs and benefits: that is, in money terms.

If both these questions can be answered satisfactorily, the CBA equation could be rewritten as follows.

$$\Sigma(B_t - C_t \pm E_t)(1+r)^{-t} > 0 \quad (4.3)$$

where E_t = environmental change (cost or benefit) at time t

To address the first question, consider the situation where a community relies on catching fish, from an unpolluted river, for its income. If a factory, sited upstream, then discharges waste directly into the river, the likelihood is that the fish population in the river will be affected and, as a direct consequence, the livelihood and welfare of the fishing community downstream will decline. In this instance, there is clearly a cost associated with the pollution of the environment – that is, the loss of income suffered by the community.

This simple example illustrates that environmental 'goods' are not free. Whenever they are used, either as a resource or as a waste depository, they do incur some form of cost, whether that be in terms of loss of income, or a reduction in health or welfare, as a consequence of air pollution, high noise levels, or polluted land or water, or in a number of other ways. The only reason that these costs are not always evaluated is because there is no market for environmental goods – we cannot buy clean air, or quiet surroundings, at least not directly. Nevertheless these things do have value to people, and they increase or decrease their welfare in much the same way that other commodities, such as a new car, a washing machine, or the provision of electricity, affect the quality of people's lives.

But how can environmental costs and benefits be measured in money terms – how can they be 'monetized'? Over the last few years, environmental economists have been developing new approaches to valuing the environment, and many of these techniques can now be used to incorporate environmental costs and benefits in traditional cost–benefit analysis.

Environmental values

Environmental economists ascribe three distinct forms of value to the environment. The first of these is what is termed *actual use value*, and this refers to those benefits directly derived by people who actually make use of the environment and gain direct benefit, such as farmers, fishermen, and hikers. Polluters also derive a benefit from the environment as a consequence of being able to dispose freely of their waste.

The second form of environmental value is *option use value*, which is the value that some people put on the environment in order to preserve it for future use – either by themselves or by future generations. For example, they may believe that a particular area of scenic beauty is worth preserving because, even if they do not use it now, they may want to use it in the future.

In addition to this, some people place a value on the environment for its mere existence, even if they see no probability that they will ever make use of that environmental asset. Some people, for example, place a value on the rainforests, or the preservation of certain endangered species, even though there is no possibility of their ever seeing them. This is the *existence value* of the environment.

The *total economic value* of an environmental asset is equal to actual use value plus the option use value plus the existence value.

Methods for valuing the environment

The main methods that have been developed for 'monetizing' environmental costs and benefits are:

- the effect on production methods
- preventive expenditure and replacement cost
- human capital
- hedonic methods
- the travel cost method (TCM)
- contingent valuation.

The effect on production methods

In its simplest form, an example of this method might be the reduction in fish catches due to pollution of a watercourse as described previously. However, changes in market prices for goods and services must also be considered, as these will influence the net cost or benefit. For example, if increased production saturates the market and prices then fall, the net benefit may be reduced.

The procedure for quantifying the effect on production is a two-stage process. First, the link between the environmental impact and the amount of lost production must be established; and second, the monetary value associated with the changes must be calculated.

It is often very difficult to establish the physical link between the affected environment and the change in output: for example, to establish the relationship between the effect of fertilizers used in an upland catchment area of a lake on diminishing fish stocks in the lake. There may be other causes, and it is often an extremely complex task to disentangle the various factors contributing to a loss in production, particularly when trying to separate man-made from natural effects.

Determining the effect requires a 'with' and a 'without' scenario to be established. This is particularly difficult, if not impossible, when project actions have already been initiated.

Despite the difficulties mentioned above, this technique is probably the most widely used environmental valuation method.

Preventive expenditure and replacement cost

This method assesses the value placed on the current environment in terms of what people will spend to preserve it or stop its degradation (defensive expenditure), or restore the environment after it has been degraded. For example, the cost of additional fertilizers that are used to compensate for the loss of natural fertility in the soil arising from a particular project could be used as the environmental cost of the project. Similarly, the cost of installing double glazing in houses affected by noise from a road or airport can be regarded as the minimum cost of this form of environmental pollution arising from the project. In some cases, of course, these costs are real and actually become part of the project cost. However, even if the preventive measures are not taken, there is a cost to those affected, and it should be given consideration in the appraisal of the project.

This is often done by estimating the cost of a shadow or compensating project, which would be required to bring the environment to the same quality that it had before the main project. An example of this might be the drilling of boreholes and the establishment of an irrigation system for people displaced, by a dam project, from a village that had good natural irrigation to an area where crop irrigation is problematic.

Human capital

This method considers people as economic units and their income as a return on investment. It focuses on the increased incidence of disease and poor health, and the consequent cost to society in terms of increased health care costs and reduced potential for earnings by the affected individuals. The method suffers from the problem that it is often very difficult to isolate a single cause of productivity loss. For example, poor health may be caused by unclean water, poor domestic hygiene, polluted air, or a wide variety of other factors.

However, if a direct causal link with a particular environmental pollutant can be proven, the environmental cost of the pollution can be calculated by assessing the loss of earnings of those people affected and the additional costs of medical care required. Although relevant statistics may be available in developed countries, it is unlikely that the method could be used in many developing countries, because of the lack of records and inadequate information.

Hedonic methods

These methods use surrogate prices for environmental 'goods' for which there are no direct markets. They seek to link environmental goods with other commodities that do have a market value. For example, the price of a three-bedroom semi-detached property will vary according to the environment in which it is located. People are prepared to pay more money for a house located in a quiet area with a pleasant view, than they would for essentially the same house located next to a noisy, dirty factory. The house market can therefore be used as a 'surrogate market' for the environment. There are of course many other factors that affect house prices – such as the proximity to schools, and transportation routes – and these factors must be isolated and removed, using techniques such as multiple regression analysis, if the method is to prove in any way reliable. The method thus requires large amounts of data, which are then very difficult to analyse. However, it has been used successfully to evaluate the costs of noise and air pollution on residential environments.

Travel cost method

This method is particularly appropriate for valuing recreational areas with no or low admission charges. The value of the site is calculated as the cost incurred by the visitors in both travelling expenses and the time spent travelling. The further people travel to the site, the greater value they place on it, as they will use more fuel, incur higher fares and spend more hours in transit to and from the site.

Surveys must be carried out at the site in question. Visitors are asked how far they have travelled, how long the journey took, what type of transport they used, whether they visited any other locations, and how often they visit the site. The results of the survey, together with information on the total number of visitors to the site each year, are used to determine an annual total cost incurred by all visitors to the site. It is assumed that this cost must be a minimum value that is placed on the amenity.

The method is mainly used to determine the value of preserving national parks and other sites of scenic, recreational, and scientific interest.

Contingent valuation method

The contingent valuation method is a technique that can be used where actual market data are lacking. The method involves asking people what their preferences are and how much they are willing to pay ('willingness to pay' – WTP) for a certain environment benefit or how much they would need to be paid ('willingness to accept' – WTA) in compensation for losing a certain environment benefit.

The technique usually involves questionnaires sent to the concerned parties, or house to house interviews. This technique may be used on its own or in conjunction with the other techniques described above.

4.12 The future of EIA

EIA is now a recognized and accepted practice, and is used widely in appraising projects, primarily at the sanction stage. In many countries it is now incorporated into the legislation relating to planning and approval, and many international funding agencies now require an EIS as part of their loan or grant approval procedures.

If EIA is to continue to develop and to maintain credibility it is essential that it is seen not just as a means to obtain a permit to develop, but as an environmental management tool to be used throughout the life of the project – from inception to design, construction, operation and decommissioning. This will only come about if monitoring and auditing are conscientiously carried out and the results fed back into the EIA process.

The valuation of environmental impacts is likely to improve as more data are gathered and confidence in the methods grows. It is already possible to incorporate some environmental impacts as costs and benefits in conventional cost–benefit analysis. The sceptics point to inaccuracies associated with the monetary evaluation of environmental impacts, but even direct costs of design and construction cannot be predicted accurately. The way ahead is to recognize that, as with direct costs, social and environmental costs can only be estimated within certain limits of confidence, and to build these ranges, rather than discrete values, into our cost–benefit appraisal calculations.

Further reading

Hughes, D. (1992) *Environmental Law*, 2nd edn, Butterworth, London.

Turner, R.K., Pearce, D. and Bateman, I. (1994) *Environmental Economics – An Elementary Introduction*, Harvester Wheatsheaf, Hemel Hempstead.

Wathern, P. (ed.) (1988) *Environmental Impact Assessment – Theory and Practice*, Routledge.

Winpenny, J.T. (1991) *Values for the Environment – A Guide to Economic Appraisal*, HMSO.

Chapter 5
Cost Estimating in Contracts and Projects

When considering a project, one of the earliest requirements is to obtain an estimate of the likely budget for the project. At the inception of the project, precise data are not always available and appropriate techniques of cost estimating have to be adopted. At each stage the promoter requires the estimate to be the most likely prediction of project cost possible. The purpose of this chapter is to indicate the theory, techniques and practical implications of project cost estimating throughout the stages of the project cycle.

5.1 Cost estimating

The record of cost management in the civil engineering construction industry is not good. Many projects show massive cost and time over-runs. These are frequently caused by underestimates rather than failures of cost management or contract administration.

Estimates of cost and time are prepared and revised at many stages throughout the development of a project or contract (Figure 5.1). They are all predictions, the best approximation that can be made, and it would be unrealistic to expect them to be accurate in the accounting sense; the objective is to predict the most likely cost of the project. The degree of realism and confidence achieved will depend on the level of definition of the work and the extent of risk and uncertainty, giving a range of most probable costs. This range can be plotted against time to give a cost envelope, as illustrated in Figure 5.2. This is idealized; for example, it assumes the project's cost evolution is continuous, which is often not the case.

Generally, such envelopes show that there is a narrowing range and increasing certainty as the project progresses, certainty as to the actual cost being achieved when the final account is settled and all project-related costs have been audited. The band is wider when the project

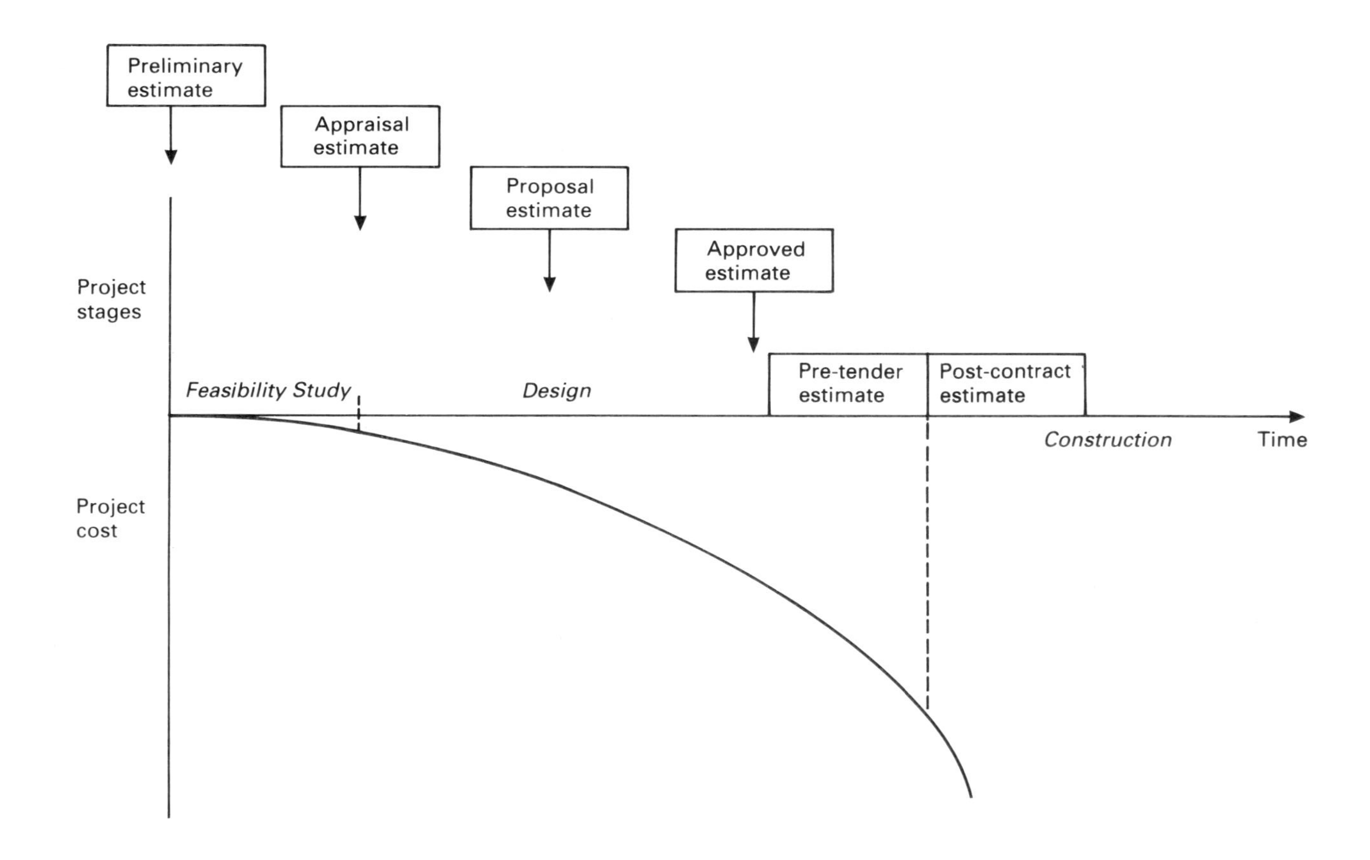

Figure 5.1 Estimates at project stages.

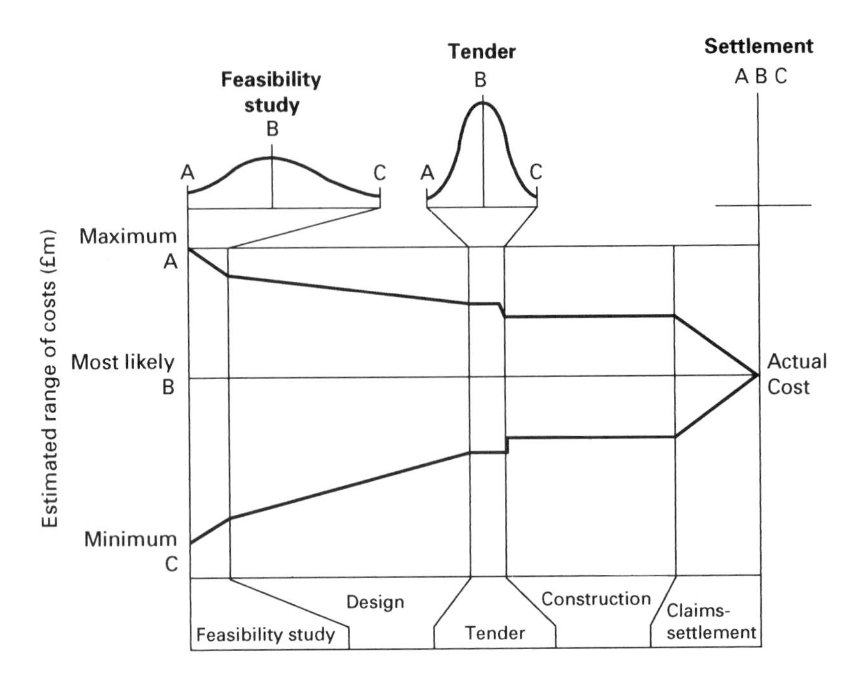

Figure 5.2 Range of estimates over the project cycle (Barnes).

commences because information is at a minimum (time and cost data, scope and organization) and many risks are latent, unrecognized by the project team, who tend not to review the evolution of the project. Project risks decrease over the life of the project, but not in a continuous way, and from time to time there may be increasing risks, or new risks that arise during the project's development.

The estimated base cost plus contingencies that are likely to be spent must be close to the 'most likely' cost from the earliest stage if the common experience of underestimation is to be avoided. It follows therefore that the 'most likely' value is closer to the 'maximum' than the 'minimum' to allow for cost growth. Any estimate should be presented as a most probable value with a tolerance; particular areas of risk and uncertainty should be noted and, if necessary, a specific contingency allowance or tolerance should be included in the estimate.

It is commonly reported that the requirements of an estimate are to predict the cost and schedule for the work, identify and quantify potential problems and risks and forecast expenditure.

The two key points at which estimates are prepared in the project life-cycle are at sanction, when the client becomes committed, and at tender, when the contractor becomes committed.

5.2 Cost and price

It is important, when reviewing estimating techniques, to be clear as to the definition of cost and price. Here they are defined in the same way as in the Overseas Development Administration Cost Estimating Guide.

Cost is the cost directly attributable to an element of work, including direct overheads: for example, supervision. Price is the cost of an element of work, plus allowance for general overheads, insurances, taxes, finance (i.e. interest charges) and profit (sometimes known as 'the contribution').

The first aim is to estimate the most probable cost of the works. The cost of an element of the works comprises quantity proportional, time-related and fixed costs. Quantity proportional costs are the direct costs of materials in the permanent works, with some exceptions: for example, the cost of concrete may vary whether or not it is batched.

Time-related costs typically relate to plant and labour. The cost of operating an excavator is a time-related cost. It needs an operator, maintenance and fuel, whatever volume of rock is excavated. If the excavator has been bought recently then there will be finance charges and depreciation to take into account. The cost of bringing the excavator to site and taking it away are lump sum start and finish costs. Payment of a specialist subcontractor is another example of a fixed cost.

These costs usually have to be translated into quantity proportional unit rates in a bill of quantities, confusing control, because if the time taken to carry out the work is longer (but the quantity remains constant) the costs will increase but the payment due will remain the same. It should also be noted that labour and plant numbers cannot be varied daily, again emphasizing the fact that the cost is related to the provision of resource, not the quantity of work completed.

These costs refer to the permanent works. Any temporary works must also be covered by the prices charged by the contractor (or subcontractor).

Finally, the direct overhead costs of management, establishment, and consumables that can be charged against a project or contract should be assessed and spread over the other cost centres, giving the total cost to perform the work.

The price of the work is derived from the cost and is sometimes described as cost plus mark-up. This does not truly reflect the elements that make up the price. The price must cover general overheads incurred by an organization which cannot be charged directly to a particular project or contract (administrative staff, senior management, cost of maintaining the head office buildings, and insurances) and the payment of taxes, interest charges on monies borrowed and allowances or contingencies for risks and uncertainties.

As a commercial organization the contractor must make a profit after tax, the amount varying depending on factors such as the type of work, the size of project or contract, the extent of competition and the desire of the organization to secure the work. These factors are often referred to as 'the state of the market'. The prediction of the price when taking these factors into account is very difficult.

There are a number of other factors that influence the overall cost and hence price of engineering works. These may include the location, the degree of innovation, the type of contract, the method of measurement, the payment conditions and the risk surrounding the project.

5.3 Importance of the early estimates

Particular care should be taken when preparing the first estimate for the project as it provides a basis against which further funds will be released, and future estimates will be compared with it. It is also at this point that the capital cost for the project will be considered as part of a full financial appraisal of the project and the decision whether or not to proceed with the project will be taken. The promoter(s) of the project should not base the sanctioning decision on the first estimate but this has to be balanced by the extra costs incurred as more detailed design is completed and more detailed estimates are undertaken.

The earliest estimates are primarily quantifications of risks. Effective estimating at this stage requires that the estimator not only has access to comprehensive historical data, and is capable of choosing and applying the most appropriate technique, but also has, in conjunction with other members of the project team, the experience to make sound judgements regarding the levels of (largely unquantifiable) risk. Projects that are in any way unusual require exercises in risk identification, risk assessment and the selection of the most appropriate response. With regard to estimating, this translates into the quantification of allowances for uncertain items, specific contingencies and general estimating tolerances. The programme should also be reviewed, and the costs of delay, the use of float or acceleration should be assessed with the appropriate contingencies included in the estimate.

This can be a time-, resource- and cost-intensive exercise but, fortunately, it is not required for all projects. The first step towards assessing the risk is to identify potentially high risk projects. This can be achieved by asking simple questions, regardless of the size, complexity, novelty or value of the particular project. The following were identified in the *Overseas Development Administration Guide*.

- Is the promoter's business or economy sensitive to the outcome of the project in terms of the quality of its product, capital cost and timely completion?

Then:

- Does the project require new technology or development of existing technology?
- Are there any major physical or logistical restraints such as extreme ground conditions or access problems?
- Does the project require a novel method of construction?
- Is the project large and/or extremely complex?
- Is there an extreme time constraint?
- Is the project's location one in which the parties involved – promoter, consultants and contractors – are likely to be inexperienced?
- Is the project sensitive to regulatory changes?
- Is the project in a developing country?

Testing new projects against such criteria is the first step towards improving the realism of initial estimates. If the project does prove to be high risk then a further, detailed risk analysis should be performed.

A list of the main sources of risk should be prepared. This may be as few as five but should not exceed fifteen. Examples of the main risk sources are: promoter; host government; funding; project definition; state of design; local conditions; construction and plant; logistics; base data; inflation; exchange rates; and *force majeure*. This list contains both qualitative and quantitative risks. The probabale impact and probability of occurrence should be assessed so that the most important factors can be identified. The estimator's attention should focus on those risks that are high impact and high probability.

5.4 Estimating techniques

The five basic estimating techniques available to meet the project needs outlined above are summarized, together with the data required for their application, in Table 5.1.

Three of the techniques (global, factorial and unit rates) rely on historical cost and price data of various kinds. There are certain points to

Table 5.1 Data for the basic estimating techniques (Smith, 1995).

	Estimating technique				
	Global	Factorial	Man-hours	Unit rate	Operational
Project data required	Size/capacity Location Completion date	List of main installed plant items Location Completion date	Quantities Location Key dates Simple method statement Completion date	Bill of quantities (at least main items) Location Completion date	Materials quantities Method statement Programme Key dates Completion date
	List of potential problems, risks, uncertainties and peculiarities of the project				
Basic estimating data required	Achieved overall costs of similar projects (adequately defined)	Established factorial estimating system Recent quotes for main plant items	Hourly rates Productivities Overheads Materials costs	Historical unit rates for similar work items Preliminaries	Labour rates and productivities Plant costs and productivities Materials costs Overhead costs
	Inflation indices Market trends General inflation forecasts	Inflation indices (for historical prices) Market trends General inflation forecasts	Hourly rate forecasts Materials costs forecasts Plant data	Inflation indices Market trends General inflation forecasts	Labour rate forecasts Materials costs forecasts Plant capital and operating costs forecasts

note when using historical data, including the following: the data sample size should be large; cost data should relate to a specific base date; the data should be updated to values that would be relevant at the base date (possibly by using an inflation index); and market effects should be accounted for.

Global

This term describes the 'broadest brush' category of technique, which relies on libraries of achieved costs of similar projects related to the overall size or capacity of the asset provided. This technique may also be known as *order of magnitude*, *rule of thumb* or *ballpark* estimating. Examples are:

- cost per megawatt capacity of power stations;
- cost per metre/km of roads/motorways;

- cost per square metre of building floor area or per cubic metre of building volume;
- cost per tonne of output for process plants.

The technique relies entirely on historical data and therefore must be used in conjunction with inflation and a judgement of the trends in levels of prices (i.e. market influence) to allow for the envisaged timing of the project.

The use of this type of 'rolled up' historical data is beset with dangers, especially inflation, as outlined generally above, but more specifically:

- Different definitions of what costs are included – for example:
  - final accounts of all contracts including settlements of claims and any *ex gratia* payments;
  - engineering fees and expenses by consultants/contractors/client, including design, construction supervision, procurement and commissioning;
  - financing costs;
  - land;
  - directly supplied plant and fittings;
  - transport costs;
  - taxes, duties, etc.
- Different definitions of measurement of the unit of capacity – for example:
  - a metre/kilometre of motorway: an overall average including pro rata costs of interchanges or estimated separately;
  - square metre of building floor area: measured inside or outside the external walls;
  - the unit with included associated infrastructure (for example, transmission links/roads for power stations);
  - cubic metre of building volume: height measured from top of ground floor or top of foundations.
- Not comparing like with like – for example:
  - different terrain and ground conditions – such as roads across flat plains compared with mountainous regions;
  - different logistics depending on site location;
  - scope of work differences, such as power stations with or without workshops and stores, housing compounds;
  - differing standards of quality, such as different runway pavement thicknesses for different levels of duty;
  - process plants on greenfield sites and on established sites;
  - item prices taken out of total contract prices (especially turnkey)

may be distorted by front end loading to improve the contractor's cash flow especially for hard currency items.

- Inflation:
  - different cost base dates – as noted above it is essential to record the 'mean' base date for the achieved cost and use appropriate indices to adjust to the forecast date required.
- Market factors:
  - competition for resources during periods of high activity;
  - developing technology may influence unit costs.

Many of these items are obvious, but for projects that do not have a continuous gestation period and may involve several different parties from time to time, it is essential that they are thoroughly checked.

A scrutiny of all these dangers, especially the effects of inflation, must be made before any reliance can be placed on a collection of data of this type. It follows that the most reliable data banks are those maintained for a specific organization, where there is confidence in the management of the data. The wider the source of the data, the greater is the risk of differences in definition.

However, as long as care is taken in the choice of data, the global technique is probably as reliable as an over-hasty estimate assembled from more detailed unit rates drawn from separate, unrelated sources and applied to 'guesstimates' of quantities.

Factorial

These techniques are typically used for process plants and power stations where the core of the project consists of major items of plant, which can be specified relatively easily and have current prices obtained from suppliers. The techniques provide factors for a comprehensive list of peripheral costs, such as pipework, electric, instruments, structure or foundations. The estimate for each peripheral will be the product of its factor and the estimate for the main plant items.

The technique does not require a detailed programme, but nevertheless one should be prepared to identify problems of construction, lead times for equipment deliveries and planning approval which will go undetected if the technique is applied in a purely arithmetical way.

An explanation of the technique for process plants is given in *A New Guide to Capital Cost Estimating* (Institution of Chemical Engineers and Association of Cost Engineers). The success of the technique depends to a large extent on four factors:

(1) The reliability of the factors, which should preferably be the result of long experience of similar projects by the estimator's organization. During periods of significant design development, certain factors can change rapidly (for example, instrumentation and controls systems).
(2) The reliability of the budget prices for the main plant items. The estimator is still required to make a judgement on the value to include in his estimate, depending on the state of the market and the firmness of the specification.
(3) The adoption of the technique as a whole so that deficiencies in some areas will compensate for excesses in others.
(4) The experience of the estimator in the use of the technique and his/her ability to make relevant judgements.

The technique has the considerable advantage of being predominantly based on current costs, thereby taking account of market conditions and placing little, if any, reliance on inflation indices.

Factorial techniques are not normally reliable for site works, including most civil and building and mechanical and electrical installation work, except in a series of projects where the site circumstances are closely similar. In most overseas locations the site works would need to be estimated separately using a more fundamental technique such as operational estimating.

Man-hours

This is probably the original estimating technique. It is most suitable for labour-intensive construction and operations such as fabrication and erection of piping, mechanical equipment, electrical installations, and instrumentation work where there exist reliable records of productivity of different trades per man-hour (for example, process plant construction and fabrication of offshore installations). The total man-hours estimated for a given operation are then costed at the current labour rates and added to the costs of materials and equipment. The advantages of working in current costs are obtained.

The technique is similar to the operational technique. However, in practice, it is often used without a detailed programme, on the assumption that the methods of construction will not vary from project to project. Experience has shown, however, that where they do vary (for example, the capacity of heavy lifting equipment available in fabrication yards), labour productivities and consequently the total cost can be affected significantly. It is recommended that a detailed programme is prepared when using this technique, to identify constraints peculiar to the project. The prediction of cash flow requires such a programme.

Unit rates

This technique is based on the traditional 'bill of quantity' approach to pricing construction work. In its most detailed form a bill of quantities will be available containing the quantities of work to be constructed, measured in accordance with an appropriate method of measurement. The estimator selects historical rates or prices for each item in the bill, using information from recent similar contracts, or published information (for example, price books for building or civil engineering), or 'built-up' rates from his analysis of the operations, plant and materials required for the measured item. As the technique relies on historical data it is subject to the general dangers outlined above.

When a detailed bill of quantities is not available, quantities will be required for the main items of work and these will be priced using 'rolled-up' rates, which take account of the associated minor items. Taken to an extreme, the cruder unit rate estimates come into the area of global estimates as described above (for example, unit rate per metre of motorway).

The technique is most appropriate to building and repetitive work where the allocation of costs to specific operations is reasonably well defined and operational risks are more manageable. Sophisticated methods of measurement for building have been developed in the UK, and internationally many contractors are able to tender by pricing bills of quantities using rates based directly on continuing experience. Nevertheless, it is essential that the rates are selected from an adequate sample of similar work with reasonably consistent levels of productivities and limited distortions arising from construction risks and uncertainties, for example access problems.

It is less appropriate for civil engineering, where the method of construction is more variable and where the uncertainties of ground conditions are more significant. It is also likely to be less successful for both civil engineering and building projects in locations where few similar projects have been completed in the past. Its success in this area depends much more on the experience of the estimator and his access to a well-understood data bank of relevant 'rolled-up' rates.

Unit rates quoted by contractors in their tenders are not necessarily directly related to the items of work they are pricing. It is common practice for a tenderer to distribute the monies included in his tender across the items in the bill to meet objectives such as cash flow and anticipated changes in volume of work, as well as taking some account of the bill item descriptions. It is unlikely that similar 'weighting' is easily carried out by all tenderers in an enquiry and therefore it is not easily detected. It follows that tendered bill unit rates are not necessarily reliable guides to prices for the work described.

From the employer's point of view the technique does not demand an examination of the programme and method of construction, and the estimate is compiled by the direct application of historical 'prices'. It therefore does not encourage an analysis of the real costs of the work of the kind that would need to be undertaken by a tendering contractor for any but the simplest of jobs. Nor does it encourage consideration of the particular peculiarities, requirements, constraints, and risks affecting the project.

There is a real danger that the precision and detail of the individual rates can generate a misplaced level of confidence in the figures. It must not be assumed that the previous work was of the same nature, carried out in identical conditions and with the same duration. The duration of the work will have a significant effect on the cost. Many construction costs are time related, as are the fees of supervisory staff, and all are affected by inflation.

It is therefore recommended that a programme embracing mobilization and construction is prepared. This should be used to produce a check estimate in simplified operational form where there is any doubt about the realism of the unit rates available.

Despite its shortcomings, unit rate estimating is probably the most frequently used technique. It can result in reliable estimates when practised by experienced estimators with good, intuitive judgement, access to a reliable, well-managed data bank of estimating data and the ability to assess the realistic programme and circumstances of the work.

Operational cost (resource cost)

This is the fundamental estimating technique, as the total cost of the work is compiled from consideration of the constituent operations or activities revealed by the method statement and programme and from the accumulated demand for resources. Labour, plant and materials are costed at current rates. The advantage of working in current costs is obtained.

The most difficult data to obtain are the productivities of labour and construction plant in the geographical location of the project and especially the circumstances of the specific activity under consideration. Claimed outputs of plant are obtainable from suppliers, but these need to be reviewed in the light of actual experience. Labour productivities will vary from site to site depending on management, organization, industrial relations, site conditions and other factors, and also from country to country. Collections of productivity information tend to be personal to the collector and indeed this type of knowledge is a significant part of the 'know-how' of a contractor and will naturally be jealously guarded.

The operational technique is particularly valuable where there are significant uncertainties and risks. Because the technique exposes the basic sources of costs, the sensitivities of the estimate to alternative assumptions/methods can be investigated easily and the reasons for variations in cost appreciated. It also provides a detailed current cost/time basis for the application of inflation forecasts and hence the compilation of a project cash flow.

In particular the operational technique for estimating holds the best chance of identifying risks of delay as it involves the preparation of a method of construction and sequential programme including an appreciation of productivities. Sensitivity analyses can be carried out to determine the most vulnerable operations, and appropriate allowances included. Action to reduce the effects of risks should be taken where possible.

It is the most reliable estimating technique for civil engineering work, and it is frequently used by major contractors and an increasing number of consulting engineers. Its execution is relatively time-consuming and resource-intensive compared with other techniques. However, estimating organizations geared up to this technique accumulate data in an operational form, which enables them to prepare even preliminary estimates with some appreciation of the more obvious risks, uncertainties and special circumstances of the project.

5.5 Suitability of estimating techniques to project stages

The objective should be to evolve a cost history of the project from inception to completion with an estimated total cash cost at each stage near to the eventual out-turn cost. This can be achieved if the rising level of definition is balanced by reducing tolerances and contingency allowances that represent uncertainty. Ideally, each estimate should be directly comparable with its predecessor in a form suitable for cost monitoring during implementation and for a usable record in a cost data bank. This may, in practice, be difficult to achieve.

There is some correlation between the five estimating techniques which have been described and the estimating stages which have been defined. This is related to the level of detail available for estimating.

Preliminary

This is an initial estimate at the earliest possible stages, there are likely to be no design data available and only a crude indication of the project size or capacity, and the estimate is likely to be of use in provisional planning of capital expenditure programmes.

At this stage the global estimating technique can be used, which is a crude system that relies upon the existence of data for similar projects assessed purely on a single characteristic such as size, capacity, or output. Widely used on process plants is the factorial method, where the key components can be easily identified and priced, and all other works are calculated as factors of these components.

Feasibility

Sometimes known as an appraisal estimate, this comprises directly comparable estimates of the alternative schemes under consideration. It should include all costs that will be charged against the project to provide the best estimate of anticipated total cost, and if it is to be used to update the initial figure in the forward budget then it must be escalated to a cash estimate.

A price can be described by the following equation:

$$\text{Price} = \text{cost estimate} + \text{risk} + \text{overheads} + \text{profit} + \text{mark-up}$$

The size of the profit margin and the commercial decision making behind the selection of the percentage mark-up or mark-down are not discussed in detail here.

The basic cost estimate is the largest of all these elements, often accounting for more than 90% of the total price. Usually the basic cost estimate is derived from the unit rate or operational assessment of the labour, plant, materials and subcontract work required. Quotations are required for materials and subcontractors. Typically materials account for between 30 and 60% of a project's value; subcontractors can typically account for between 20 and 40%.

The cost of the company's own labour is usually calculated per hour, per shift or per week. The cost to the company of employing its own labour is greater than that paid directly to the employee. The elements in the calculation could include such factors as plus rates for additional duties, tool money, travel monies, National Insurance, training levies, employers' and public liability insurance and allowances for supervision.

Plant may be obtained for a contract either internally or externally. Quotations for the plant required are therefore obtained from external hirers or from the company's own plant department. It is rare for UK contractors in the domestic market to calculate the all-in plant rate from first principles. This calculation is usually undertaken by the plant hire company.

Overheads (or on-costs) for the project could include allowances for site management and supervision, clerical staff and general employees, accommodation, general items and sundry requirements.

Design

This is an estimate for the selected scheme. It usually evolves from a conceptual design until immediate pre-tender definitive design is completed.

A man-hours method is most suitable for labour-intensive operations, like design, maintenance or mechanical erection, and work is estimated in total man-hours and costed in conjunction with plant and material costs.

The decisions made in design on the size, quality and complexity of a project are the greatest influence on the final capital costs of a project. As a design develops more and more capital cost is committed on behalf of the client until at tender stage with the design complete, or virtually complete, the client is committed to a high percentage of the tender value. Unless a redesign is undertaken, with the consequent loss of fees and time, the ability to save money while maintaining the original design concept is very limited. The design budget estimate should confirm the appraisal estimate and set the cost limit for the capital cost of the project.

Construction

This is a further refinement to reflect the prices in the contract awarded. This would require some redistribution of the money, for example in the bill of quantities in an admeasurement contract, and assist more efficient management of the contract.

The unit rate method is a technique based on the traditional bill of quantities approach where the quantities of work are defined and measured in accordance with a standard method of measurement.

5.6 Estimating for process plants

The most significant difference between estimating for a process plant project and estimating for other types of engineering project is that the base cost of the process estimate is derived from material and equipment suppliers, plant vendors, specialist contractors and subcontractors. These components commonly account for about 80% of the total cost of the project. Consequently, the project estimator must ensure that firm quotations, together with guaranteed delivery dates and installation schedules, are confirmed with all suppliers and subcontractors.

The main process plant contractor, the engineering contractor, usually carries out the design, procurement and management functions, which accounts for most of the remaining cost about 20%. The first task is for the engineering contractor to estimate his own base costs as accurately as possible, for the work to be undertaken at the detailed design, procurement, project management and site supervision stages of the project. Typically the construction of process plants is labour intensive and hence this assessment is made using the man-hour estimating technique.

The engineering contractor's control philosophy must be aimed so as to minimize change at the detailed engineering phase. The initial global estimate can be made by comparative means with similar types of plant based on throughput. The estimate will have considered the principal quantities of the work, the items to be subcontracted, the materials and plant for which quotes are required, critical dates for actions by subcontractors and suppliers, and whether any design alternatives should be pursued.

At this stage the effects of layout and location are considered by the preparation of the base, or net cost estimate. Costs will be established for project and engineering services, which is usually done in-house, and any remaining quotes are substantiated and confirmed. It is important to have a good definition of the scheme, including general arrangement, piping and instrument diagrams, equipment lists and material specifications.

In the production of a net cost, allowances and contingencies need to be considered. Allowances can be defined as costs added to individual estimated costs to compensate for a known shortfall in data, or to provide for anticipated developments. Often, when firm quotes are replaced with contracts and orders after the successful tender bid, more detail is available than was the case earlier. Contingency is an adjustment to the net cost and has to be considered no matter how much detailed work has been completed during the preparation of the estimate, as there will undoubtedly be uncertainties that will affect the final cost.

Contingency adjustments allow for the unknown element and also for any factors outside the control of the engineering contractor that are perceived as likely to affect the final cost.

The cost of preparing a fixed price estimate can be as high as 1–3% of total cost, compared with 0.5–1% for building and civil engineering work, and take up to six months to prepare. Typically a three-month tender period for most major process plant projects is allowed by most clients, which is really too short to enable all the requirements of the estimate to be completed.

5.7 Information technology in estimating

The estimating techniques described were developed mainly as manual methods. The development of computer hardware, dedicated computer software packages and increased computer literacy among the professionals engaged in project management and estimating has resulted in the application of information technology (IT) tools to facilitate and assist in the estimating process.

As in manual estimating, the role of computer estimating will depend upon the user's requirements. Those of a client at the design stage will be very different from those of a contractor at tender. The data available for the production of an estimate will also be different. This section does not contain details of specific hardware or software products, which are well marketed and are almost continually updated and improved; rather it focuses on some of the main functions that IT can adopt in this process.

IT is concerned with the storage, transfer and retrieval of data, manipulation and calculation using data, and presentation of output. The degree of complexity will vary, as will the operation of the user interface, ranging from data libraries to fully automated expert systems. Most computer software estimating packages incorporate a data library for storing data, a range of methods for manipulating direct costs to produce a price, the ability to update or alter any of the input data, and appropriate reporting formats, often consistent with standard methods of measurement documents.

The software packages attempt to be user-friendly by permitting data input or price build-up by the estimator to resemble closely the normal way of working. The estimator has to spend less time performing routine calculations and arithmetical checks while the basic method remains unchanged, but has more time to use judgement and experience where appropriate. It is important to note that a computer-based estimate is only as accurate as the input data and that the use of a computer in itself will not necessarily increase the accuracy of estimates. Indeed, if the data library is not kept up to date or if it is applied without careful thought then the estimate produced might suffer from a decrease in accuracy.

The application of IT cannot replace the role of the estimator in the project, nor is it intended to do so. Software packages are readily available to assist with most types of estimating technique, but there are also other advantages for the project manager. One of the recognized sources of errors and misunderstanding in project management has been identified as the linking interface between feasibility study, project estimate and detailed planning. Often these stages were tackled by different people using different estimating methods with associated differing assumptions, and the previous estimate was usually abandoned once that phase of the project had been completed. Current estimating software packages can be related to a single database and also be linked with other computer programs. This allows all the work done at feasibility stage to be refined, modified and developed as the project progresses, hence retaining all the information within the system. Typical interfaces relate to measurement and valuation, to bills of quantities, to the valuation of variations, and to delays and disruptions. An estimating package can be linked with a planning package, and data automatically transferred facilitating the understanding of the relationship between time and money for the project.

Expert systems are attracting considerable interest as potential aids to decision making. In an advisory role they could provide the necessary assistance to produce the cheapest cost estimate, by allowing automated decision making to select resources to match different work loads. The 'expert' approach purports to provide tools for the effective representation and use of knowledge developed from experience, enabling the optimization of resource selection. A computer simulation then begins with a preliminary questions module in which the user responds to a set of questions posed by the system.

Estimating systems are evolving within an integrated environment of construction management systems. A knowledge-based system should also develop within this environment as a decision making or decision suggesting aid.

5.8 Realism of estimates

The use of the word 'realism' in this context, rather than 'accuracy', is important. As noted above, estimates are not accurate in the accounting sense, and the make-up of the total must be expected to change.

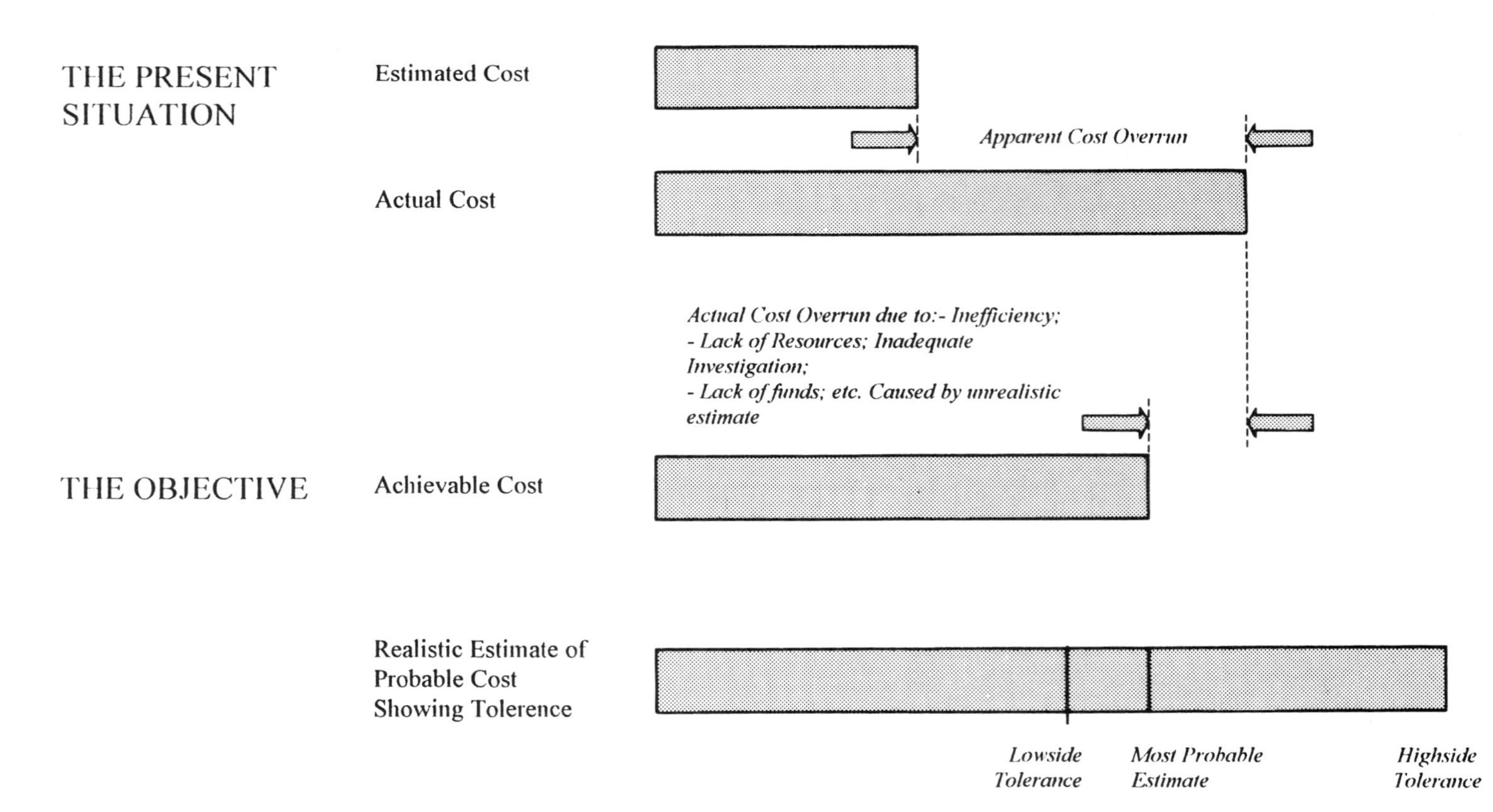

Figure 5.3 Estimating performance and objectives.

The realism of estimates will depend greatly on the nature and location of the work, the level of definition of the project or contract, and particularly on the extent of the residual risk and uncertainty at the time, as discussed above.

Studies have determined that a standard deviation of 7% was common for contractors' bid estimates in the UK process plant industry, but performance of particular companies varied from 4 to 15%

However, the ranges of accuracy for high-risk projects, and in particular development projects, may be much greater.

The variation in individual items within the estimate is much greater; consequently any system of cost control based on comparison of actual/estimated cost must be of dubious value. It is preferable to use performance measurement systems based on the integration of cost and programme. These systems are based on the concept of *earned value*.

Simulation studies suggest that improvement in estimating accuracy will produce a corresponding improvement in company performance.

Many estimating problems can be addressed by adopting:

- a structured approach
- choice of the appropriate technique
- use of the most reliable data
- consideration of the risks.

The improvement in establishing performance that should be obtained is illustrated in Figure 5.3.

Further reading

A Guide to Cost Estimating for Overseas Construction Projects (1989) Project Management Group UMIST for the Overseas Development Administration.

A New Guide to Capital Cost Estimating (1977) Institution of Chemical Engineers, Rugby.

Association of Cost Engineers (1984) *Standard Method of Measurement for Industrial Engineering Construction*, The Association of Cost Engineers, London.

Association of Cost Engineers (1991) *Estimating Checklist for Capital Projects*, 2nd edn, The Association of Cost Engineers, London.

Barnes, N.M.L. (1974) Financial control of construction. In: *Control of Engineering Projects* (ed. S.H. Wearne) Chapter 6, pp. 122–140. Edward Arnold, London.

Bower, D., Thompson, P.A., McGowan, P. and Horner, R.M.W. (1993) Integrating project time with cost and price data. In: *Developments in Civil and*

Construction Engineering Computing, The 5th International Conference on Civil and Structural Engineering Computing, pp. 41–49. Civ-Comp Press, Edinburgh.

Harris, F. and McCaffer, R. (1995) *Modern Construction Management*, 4th edn, Blackwell Science, Oxford.

McCaffer, R. and Baldwin, A.N. (1991) *Estimating and Tendering for Civil Engineering Works*, 2nd edn, Blackwell Science, Oxford.

Perry, J.G. and Thompson, P.A. (1992) *Engineering Construction Risks – A Guide to Project Risk Analysis and Risk Management*, Thomas Telford, London.

Smith, N.J. (ed.) (1995) *Project Cost Estimating*, Thomas Telford, London.

Chapter 6
Project Finance

In the UK, major infrastructure projects in the transportation and energy sectors have traditionally been funded by the public sector, while industrial and commercial projects have tended to adopt private financing. With increasing privatization these distinctions have become less clear and many projects are funded by combinations of public and private sector funding. In this chapter the sources of finance, methods of project financing, types of loans and loans in a mixture of currencies are described. Finally, the procedures for the assessment and management of financial risks and the appraisal of finance for projects are presented.

6.1 Funding for projects

Public finance has provided most of the major infrastructure projects procured in the UK and overseas in the last 50 years. Monies raised from taxation have provided all or part of the finance required for projects such as motorways, bridges and tunnels, transport systems and water and power plants. In many cases low interest rate loans or deferred or subordinated loans were provided by government in conjunction with loans from the private sector for major infrastructure projects, resulting in a hybrid, mixed funding finance package. The fact that no money is free, even to governments who borrow from the private sector and bear the costs of collecting monies themselves, has resulted in private finance being considered for a number of major projects normally procured with public funds.

The use of private finance instead of public finance for a particular project is only justified if it provides a more cost-effective solution. The financial plan of a project will often have a greater impact on its success than the physical design or construction costs. The cost of finance and its associated components are often determined by the method of revenue generation, the type of project and its location.

Often government support to lenders of project finance, in the form of guarantees, has been sufficient to ensure projects have been completed that would not have been commercially viable without such support.

6.2 Sources of finance

In major projects finance is often provided by a lender in the form of a commercial bank, a pension fund, an insurance company, an export credit agency or a development bank. Other project finance sources include institutional investors, large corporations, investment banks, niche banks and developers, utility subsidiaries, and vendors and contractors. The more attractive, in terms of potential returns, the project is, the more sources of finance it is likely to attract.

Figure 6.1 shows a number of sources of funding for projects.

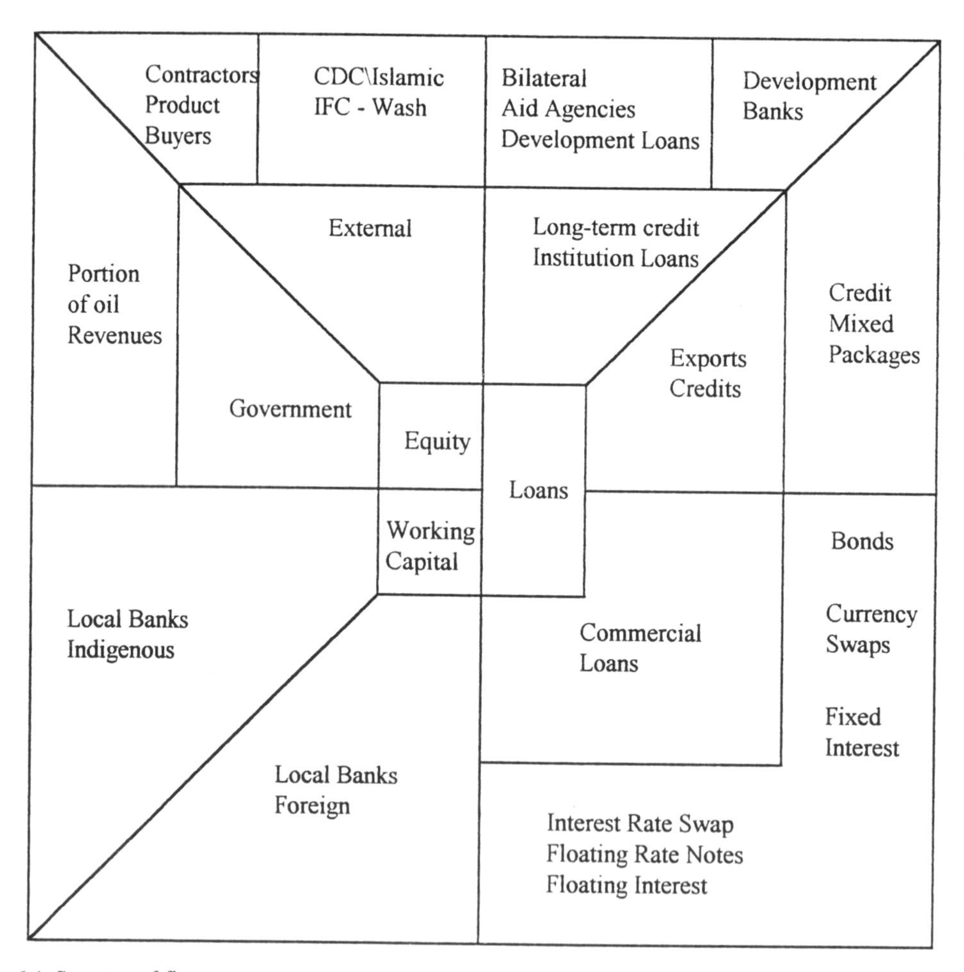

Figure 6.1 Sources of finance.

In the Channel Tunnel project, for example, the debt/equity ratio was initially set at 80/20, with finance raised from the following sources:

- equity
  - banks and contractors: founder shareholders
  - private institutions: first tranche
  - public investors: second tranche
  - public investors: third tranche
  - public investors: fourth tranche
- debt
  - commercial banks: main facility
  - commercial banks: standby facility.

In order to pay the loan arrangements, revenue is obtained from three sources: shuttle services, British and French railways, and other associated facilities. The original financing structure of this project involved 210 lending organizations, with 30% of the loans arranged through Japanese banks.

In some cases, where no equity finance is deemed necessary, finance will be provided purely on the basis of debt and normally arranged by one lender acting as a lead lender for a number of banks or institutions. In many cases governments or multinational guarantee agencies may guarantee loans over specific lending periods for specific types of project. Many lender organizations will only lend to projects with reasonably assured cash flows and sufficient government support.

6.3 Project finance

Project finance is the term used to describe the financing of a particular legal entity whose cash flows and revenues will be accepted by the lender as a source of funds from which the loan will be repaid. Thus the project's assets, contracts, economics and cash flows are segregated from its sponsors such that it is strictly limited recourse (i.e. projection) in that lenders assume some of the risk of the commercial success or failure of the project. In summary, project finance may be defined as:

> the financing of a particular economic unit in which the lender is satisfied to look initially to the cash flows and earnings of that unit as the source of funds from which the loan will be repaid and to the assets of the economic unit collateral for the loan.

Project finance provides no recourse and if project revenues are not sufficient to cover debt service, lenders have no claim against the owner beyond the assets of the project; the project, in effect, is self-funding and self-liquidating in terms of financing.

Unlike traditional public sector projects, whose capital costs are largely financed by loans raised by government, privately financed projects are financed by a combination of debt and equity capital. The ratio between these two types of capital varies between projects.

Financing a project is often the most difficult operation, with the largest risks occurring during the construction phase. Often equity finance is arranged to overcome this problem before revenue is generated by the project. This is similar to the conditions for the original loan facility for the Channel Tunnel Project, under which 70% of the project equity had to be spent before any loans could be taken up by the promoter. The 'offer of sale' of equity shares made no mention of shareholders' entitlements after the concession is transferred to the UK and French governments in July 2042. This is characteristic of investment equity capital, as unlike other forms of capital market lending the loan never comes to maturity; that is, the loan is permanent and never repaid. However, equity shares are negotiable and may be transacted on the Stock Exchange.

6.4 Debt financing contract

The contract between the sponsor and lender can only be determined when the lender has sufficient information to assess the viability of a project. In construction projects the lender will look to the project itself as a source of repayment rather than the assets of the project. The key parameters to be considered by lenders to include the following.

- *Total size of the project.* The size of the project determining the amount of money required and the effort needed to raise the capital, internal rate of return on the project and equity.
- *Break-even dates.* Critical dates when equity investors see a return on their investments.
- *Milestones.* Significant dates related to the financing of the project.
- *Loan summary.* The true cost of each loan, the amount drawn and the year in which drawdowns reach their maximum.

A properly structured financial loan package should achieve the following basic objectives:

- *Maximize long-term debt.* Allow the project entity itself to incur the debt, which would not effect the balance sheet of the sponsor's parent company.
- *Maximize fixed-rate financing.* The utilization of long-term export credit facilities or subordinated loans with low interest rates over long terms will reduce project risks.
- *Minimize refinancing risk.* Cost overruns present additional problems to a project; therefore standby credit facilities from lenders and additional capital from promoters should be made available.

In order to manage various types of risk a lender will often prepare a *term sheet*. This term sheet, based on agreement among the project participants, defines the rights and obligations of lenders, describes default conditions and remedies, and serves as the bid document for accessing capital markets.

The components of a term sheet determine whether a loan would be sanctioned for a particular project and provide a mechanism for organizations evaluating the financial parameters of a bid. The parameters may include: bank loan, index or coupon bonds, drawdown of loans, equity, dividends, preferred share, interest during construction, interest during operation, lenders fees, internal rate of return (IRR), net present value (NPV), coverage ratio, payback period of loans, standby facility, working capital and debt service ratio.

Lenders will need to be satisfied that the project can be seen to be viable for each risk analysis. In assessing the adequacy of projected cash flows the major criteria may be:

(1) *the debt/service ratio* (defined as annual cash flow available for debt service) divided by debt service; and
(2) the *coverage ratio* (defined as NPV of future after-tax cash flows over either the project life or the loan life) divided by the loan balance outstanding

While the debt/service ratio is used in the evaluation of limited resource financing in all types of industries, the coverage ratio is a well-established criterion for the evaluation of oil and gas projects.

For example, in a project where the coverage ratio is 1, the debt will be repaid with no margin of error in the cash flow projections. Ratios less than 1 mean that debt will not be repaid over the term of the calculation, and ratios in excess of 1 provide a measure of comfort should variations from the assumptions occur. The term sheet may be used to express such ratios for one or a number of risk analyses, and should the ratio be less than the lender's requirement then the loan package may be considered commercially unviable for the risks to be covered.

6.5 Types of loan

The loan structure of a finance package is often the most important ingredient to the success of a project. The structure of the loan may be in the form of *debt* (loans and debentures) or *equity finance*.

Debt: loans and debentures

The conditions of loan finance will depend on the criteria of the lender and the sponsor and the type of project considered. The main features that will need to be agreed are the repayment method, the interest rate and the security.

Repayment method

- *Mortgage.* This may be in the form of a mortgage with repayments calculated so that the total amount paid at each instalment is constant while capital and interest vary. This means that the proportion of capital repaid is initially small but progressively increases throughout the project. This form of loan is suitable for projects where no revenue is generated until the project is operational.
- *Equal instalments of principal.* The amount of principal repaid is constant with each payment, with the total amount paid decreasing over time as it consists of a constant principal repayment plus interest on outstanding principal.
- *Maturity.* The principal is repaid at the end of the loan period in one sum. This form of loan is most suitable for projects that generate a large capital sum on completion, and could only be considered if the sponsor intended to sell the facility immediately after commissioning.

Other variations on repayment structures may include a moratorium on capital repayments or interest payments for a period at the start of the loan, a bullet payment or a sunset payment at the end of the loan period. This structure is useful for projects where the project does not begin to generate revenue until fully operational.

Interest rate

The interest rate will often be governed by the money market rates, which vary considerably. Interest rates may be fixed for the period of the loan or expressed as a percentage of the standard base rate. In some cases the sponsor may negotiate a floating rate where an upper and lower limit is set on the interest rate. These floating upper and lower rates are often fixed to maximum and minimum rates by introducing a cap and a collar respectively. The sponsor is protected from interest rate increases over the agreed cap and the lender from decreases below the agreed collar.

In some cases the rate of interest will be determined as:

rate of interest = relevant margin + Libor

where Libor is the arithmetic mean of the rates quoted by the reference banks.

An example of applicable margins is:

- main loan facility
  - pre-completion 1.25%
  - post-completion 1.00%
- standby facility
  - pre-completion 1.5%
  - post-completion 1.25%
- working capital facility: 1.5%.

Security

On conventional loan finance the lender seeks to limit risk by insisting that the borrower provides security for the loan: this is usually in the form of a charge on the project's assets or through a guarantee. In projects where the amount involved may be too large for participants to provide a guarantee, non-recourse loans are normally adopted. Under this form of loan the lender accepts that the surplus cash flow over the operating costs will be sufficient to cover interest payments arising from the debt. In some cases the facility itself may have a realizable capital value.

Another type of finance that may be considered for projects, especially in developing countries, is countertrade. Under this form of agreement goods are exchanged for goods. An example of this form of agreement is the use of oil used as payment for goods and services supplied. In such a project the oil would be sold on the open market or under a sales contract to generate revenues for repayments of capital and interest.

Debentures

A debenture is a document issued by a company in exchange for money lent to the company. The company agrees to pay the lender a stated rate of interest and also repay or redeem the principal at some future date. Debenture can be traded in the same way as shares. The interest paid to debenture holders is deductible when calculating taxable profits, unlike dividend payments.

Other forms of lending may include export credits, floating interest rate loans, currency swaps and revenue bonds.

In most cases the success of projects depends entirely on cash flow. *Mezzanine finance* is an innovative form of finance to enhance return to shareholders. Mezzanine finance may be in the form of fixed or floating rate loans with second charge assets; these are similar to conventional loans in that they provide for the payment of interest and principal through a flexible amortization schedule.

Alternative finance packages may be based on:

- *royalty agreements:* agreement providing the lender with an agreed percentage of future revenues;
- *unsecured loan stock:* fixed-interest loan stock which gives the lender the right to a fixed return and to obtain repayment of the principal at the end of a stated period;
- *convertible unsecured loan stock:* fixed-interest loan stock with the right to convert to ordinary share capital at a future date;
- *redeemable preference shares:* shares giving the investor the right to a fixed return and to obtain repayment of the investment at the end of a stated period;
- *convertible preference shares:* shares giving the investor the right to a fixed return and to convert to ordinary share capital at a future date.

A typical project loan facility may include:

(1) main bank facility
(2) letter of credit facility
(3) standby facility.

Facilities (1) and (2) are available to finance any cost associated with the acquisition, operation and maintenance and costs associated with an existing facility included under the concession contract.

Facility (3) will be available to meet required drawdowns in excess of the main bank facility due to higher interest rates, lower than expected revenues in the early phase of operation and for additional works performed under the construction contract.

In addition to the loan facilities a typical finance package may also include the payment of fees such as:

- *underwriting fee:* payable as a percentage of the bank loan facility;
- *management/front end fee:* payable as a percentage of the total facilities put in place;
- *commitment fee:* calculated on the undrawn balance;
- *agency fee:* an annual fee paid to the lead lending bank;
- *success fee:* payable as a percentage of the total loan once all loans have been secured.

Equity finance

Equity finance is usually an injection of risk capital into a company or venture. Providers of equity are compensated with dividends from profits if a company or venture is successful, but no return should the venture be loss making. Debt service in the majority of cases takes first call on profits whether or not profits have been generated; dividends are paid after debt claims have been met. In the event of a company or venture becoming insolvent, equity investors rank last in the order of repayment, and may lose their investment.

The amount of equity provided is considered as the balance of the loan required to finance the project. The total finance package is often described in terms of the debt/equity ratio. In projects considered to have a large degree of risk then a larger proportion of equity is normally provided.

The advantages of an equity investment are that equity may be used as a balancing item to accommodate fixed repayments, and that equity investors are often committed to the success of a project, being organizations involved in the realisation of the project. Providers of equity fall into two categories: those with an interest in the project (contractors, vendors, operators), and pure equity investors (shareholders).

Sources of equity include: public share issue, financial institutions such as pension funds, companies and individuals, participants such as constructors, suppliers, operators, vendors and government, and international agencies such as IFC, the European Bank for Reconstruction and Development (EBRD) and the European Investment Bank (EIB).

In a number of overseas projects, finance is often provided in a mixture of currencies, often based at the time of the loan to the host country currency. Host country inflation may often result in additional loans being required to complete a project if no fixed exchange rate is agreed.

Fluctuations between the values of the currencies provided in a project loan need to be addressed by the lender.

Similarly, loans to countries that do not have transferable or exchangeable currencies, such as many eastern European and African countries, need to be based either on guarantees from a third party or on countertrade. In the case of the latter a product is provided in lieu of the loan repayments; this may be in the form of minerals or services. However, should the selling price of such minerals or services reduce during the repayment period then the lender may not even recoup the principal loaned. Lenders will often overcome this commercial risk by entering into a sales contract for the product or service at prices suitable to meet the loan schedule.

6.6 Appraisal and validity of financing projects

The financial viability of a project over its life must be clearly demonstrable to potential equity investors and lending organizations. In assessing the attractiveness of a financial package project, sponsors should examine the following elements:

- interest rate, debt/equity ratio (percentage being financed);
- repayment period, currency of payment, associated charges (legal, management and syndication fees), securities (guarantees from lenders) and documentation (required for application, activation and drawdown of loans).

Three basic financial criteria need to be achieved in projects. Finance must be cost effective, so far as possible; the skilled use of finance at fixed rates to minimize risks should be adopted; and finance is required over a long term.

The project must have clear and defined revenues that will be sufficient to service principal and interest payments on the project debt over the term of the loans and provide a return on equity that is commensurate with development and long-term project risk taken by equity investors. The EIB will normally fund infrastructure projects for a period of 25 years and industrial/process plants up to 14 years. Institutional investors such as insurance companies and pension funds often consider projects with fixed rates of return up to 20 years to match the cash flow characteristics of their liabilities. Lenders often refer to a robust finance package as one that will allow repayment of loans should interest rates increase during the operation period.

The commercial and financial considerations rather than the technical elements are normally the determinants of a successful project. The political risk, which in turn influences the financial risk, is less controllable than the technical risks, which are often allocated to the constructor or operator.

When selecting the sources and forms of capital required the following should be considered: the strength of the security package, perception of the country risks and limits, sophistication of local capital markets.

It is possible that project finance is provided entirely from debt, as in the Dartford River Crossing project. In this case the lenders take the risk that revenues will be sufficient to pay off the debt by the end of the concession period. In this project, for example, the promoter provided pinpoint equity of only £1000 (a form of equity under which shareholders do not receive dividends) and the remaining finance through loans.

In some cases government organizations may take an equity participation in the project; however, many sponsors believe the inclusion of government among the project company shareholders can lead to bureaucratic interference with project development and operation, which privatization is supposed to avoid.

In build-own-operate-transfer (BOOT) project strategies, for example, the providers of finance are compensated solely from the project revenue with recourse to the revenue stream. The payment of revenues in BOOT projects may be arranged according to the time period revenue streams or user revenue streams.

A time period revenue stream is normally a predetermined payment by the promoter to the sponsor independent of the usage of the facility but guaranteeing the promoter a secured income. For example, a promoter may lease from a sponsor a tunnel based on an annual payment irrespective of the number of vehicles using the facility.

However, in most cases user revenue streams are related to the level of use of the facility; the number of vehicles using a tunnel multiplied by the toll rate would be considered as a user revenue.

One of the most important elements in project finance is how to provide security to non-recourse or limited recourse lenders. If a sponsor defaults under a particular project strategy utilizing a non-recourse finance package, the lender may be left with a partly completed facility, which has no market value. To protect lenders, therefore, various security devices are often included to protect the lender. These may include the following.

- Revenues are collected in one or more escrow accounts maintained by an escrow agent independent of the sponsor company.

- The benefits of various contracts entered into by the sponsor, such as construction contract, performance bonds, supplier warranties and insurance proceeds, will normally be assigned to a trustee for the benefit of the lender.
- Lenders may insist upon the right to take over the project in the case of financial or technical default prior to bankruptcy and bring in new contractors, suppliers or operators to complete the project.
- Lenders and export credit agencies may insist on measures of government support such as standby subordinated loan facilities, which are functionally almost equivalent to sovereign guarantees.

In summary, the successful elements required in funding projects should include: limited and non-recourse credit, debt financing entirely in local currency, equity finance in currencies considered relatively strong, major innovations in project financing, project creditors are confident and governments accept some project risks and provide limited resources.

6.7 Financial risks

The identification of risks associated with any project is a necessary step before analysis and allocation, especially in the early stage of project appraisal. Lenders and investors will only be attracted to projects that provide suitable returns on the capital invested.

A project often has a number of risks: identifiable risks, which are within the control of one or more of the parties to the project; risks which may not be within any party's reasonable control, but may be insurable at a cost; and uninsurable risks.

By identifying risks at the appraisal stage of a project a realistic estimate of the duration and final costs and revenues of a project may be determined.

Financial risks

Financial risks are associated with the mechanics of raising and delivering finance and the availability of adequate working capital.

Financial risks may be summarized as:

Interest: type of rate, fixed, floating or capped, changes in interest rate, existing rates.

Payback: loan period, fixed payments, cash flow milestones, discount rates, rate of return, scheduling of payments.

Loan: type and source of loan, availability of loan, cost of servicing loan, default by lender, standby loan facility, debt/equity ratio, holding period, existing debt, covenants.

Equity: institutional support, take-up of shares, type of equity offered.

Dividends: time and amounts of dividend payments.

Currencies: currencies of loan, ratio of local/base currencies, mixed currencies.

Revenue risks

The risks associated with revenue generation are often considered on the basis of meeting demands and may include:

Demand: accuracy of demand and growth data, ability to meet increase in demand, demand over concession period, demand associated with existing facilities.

Toll: market-led or contract-led revenues, shadow tolls, toll level, currencies of revenue, tariff variation formula, regulated tolls, take or pay payments.

Developments: changes in revenue streams from associated developments.

Commercial risks

Commercial risks are considered when determining the commercial viability of a project. Commercial risks affect the market and revenue streams. They may include risks associated with access to new markets, size of existing markets, pricing strategy and demand. Commercial risks can arise from deterioration of a competitive market position and overly optimistic appraisals of the value of pledged securities, such as oil and gas reserves. Project revenues can either be market-tied or contract-tied; market-tied revenues normally impose higher risks on the project sponsor. Demand risks are normally uncontrollable on a road project and if demand is less than predicted then debt service may become impossible. Market risks such as failure of customers to meet sales agreements, price fluctuations, inaccurate assessment of outlets, market uncertainty, obsolescence of the product and cancellation of long-term purchasing contracts may result in commercial failure of a project. One way to minimize product market risk is direct financial participation by customers.

Market risks prior to completion of a project would normally include the following.

- Raw materials are not available when required during the construction phase.
- The market price of raw materials increases during the construction phase.
- The market price of the project's product falls or fails to rise as anticipated during the construction phase.
- Market forces change leaving no receptive market for the project's product once production commences.

In international projects, foreign exchange risks may be minimized by financing a project in the same mixture and proportion of currencies as those anticipated from the revenue streams. Cash flows determined by a number of currencies may be simplified by classifying currencies into groups. The two major base currencies used in this simplification are the US dollar and the German mark.

Commercial risks may be summarized as:

Market: changes in demand for facility or product, escalation of costs of raw materials and consumables, recession, economic downturn, quality of product, social acceptability of user pay policy, marketing of product and consumer resistance to tolls.

Reservoir: changes in input source.

Currency: convertibility of revenue currencies, fluctuation in exchange rates, devaluation.

Lenders and investors should evaluate the above risks to determine the commercial viability of a project. Should a project seem to be commercially viable from a financial point of view, then additional risk areas associated with construction, operation and maintenance, political, legal and environmental should be identified and appraised:

Sensitivity: location of project, existing environmental constraints, impending environmental changes.

Impact: effect of pressure groups, external factors effecting operation, effect of environmental impact, changes in environmental consent.

Ecological: changes in ecology during concession period.

A typical project finance analysis is summarized as:

Financial market analysis:	considers data regarding the availability, cost and conditions of financing a project
Cost analysis:	estimates the development, construction and operating costs and establishes a minimum cost of the project
Market analysis:	forecasts demand and establishes a maximum price and evaluates the commercial viability of the project
Financial analysis:	compares the cost, market and financial market analysis and establishes the relationship between costs and revenues.

A number of practical methods of reducing risks in international investment analysis may include:

- reducing the minimum payback period;
- raising the required rate of return of the project investment;
- adjusting cash flows for the cost of risk reduction;
- adjusting cash flows to reflect the specific impact of a particular risk.

Allocation of risks

A typical response by lenders to minimize risks in projects may be summarized as follows.

- *Completion risk.* Cover by a fixed price, firm date, turnkey construction contract with stipulated liquidated damages.
- *Performance and operating risk.* Cover by warranties from the constructor and equipment suppliers and performance guarantees in the operation and maintenance contract.
- *Cash flow risk.* Cover by utilizing escrow arrangements to cover forward debt service and guard against possible interruptions and take out commercial insurance.
- *Inflation and foreign exchange risk.* Cover by government guarantees regarding tariff adjustment formula, minimum revenue agreements and guarantees on convertibility at certain agreed exchange rates.
- *Insurable risks.* Cover by form of insurance such as policy to cover cash flow shortfalls mainly during the pre-completion phase of a project.
- *Uninsurable risks.* Cover by insisting host government provide some form of coverage for uninsurable risks such as *force majeure*.

- *Political risks.* Cover by political risk insurance from export credit agencies or multilateral investment agencies.
- *Commercial risk.* Cover by insurance policies such as those of the Export Guarantee Credit Department of the Department of Trade and Industry.

6.8 Financial guarantees

Guarantees required by client organizations from a sponsor may include:

Tender guarantee:	this is often requested to eliminate sponsors who do not have sufficient financial resources.
Performance guarantee:	to ensure the payment of a financial penalty should the sponsor fail to perform.
Completion guarantee:	to ensure the project is completed on time such that revenue generation can proceed.

Guarantees are normally provided by banks in the UK, insurance companies in Australia and surety companies in the USA.

In a number of projects the sponsor may be part of a larger company or in a group of companies, and it is often a requirement of a promoter organization that the parent company guarantee performance by the sponsor. Similarly, lending organizations must look to guarantees to cover such risks on loans provided to sponsors.

A private finance package is dependent on the type of project, the currencies of loan, the loan schedule and the possible effect of associated risks. If private finance is to become a major feature in procuring infrastructure projects, normally undertaken by the public sector, it is important that each project and hence finance package is considered on an individual basis. Government support must be given to projects that would not be commercially viable in the private sector to ensure the social and economic benefits of projects are enjoyed by the users.

Further reading

Jackson, M. and Richardson, M. (1990) Financing construction projects. In: *Financial Control* (ed. M. Barnes), pp. 61–62. Engineering Management Guides, Thomas Telford, London.

NEDO (1990) *Private Participation in Infrastructure*, National Economic Development Council, London.

Chapter 7
Project Cash Flow

Projects and contracts are commercial ventures. Both the promoter of a project and a contractor employed by him/her are investing money and taking financial risks in order to achieve some desired benefit or return. Project and contract management is concerned with the control of both investment and risk with the aim of achieving this return. Economic or financial evaluation of cash flows is the primary basis for decisions relating to the choice, magnitude and pattern of investment.

7.1 Cash flow

In order to quantify both the demand for money to meet the project or contract costs and the pattern of income it will generate, it is necessary to predict the cash flow. A cash flow is a financial model of the project or contract and even in its simplest basic form will provide vital information for the manager.

We are concerned with the flow of money in and out of the account per time period. Income is positive and expenditure negative: the net cash flow is therefore the difference between cash in and cash out. In most cases we aim to produce the cumulative cash flow diagram of the form shown earlier in Chapter 2.

The model should be built up in the following stages to aid thorough understanding of the details of the investment as the basic cash flow is adjusted and progressively refined. Adherence to this structure, shown diagrammatically in Figure 7.1, will aid perception of all the implications of the investment. In many cases the desired decision may be made without completing all the stages.

(1) Compile the base-case cash flow simply by adding the costs and revenues to each activity on a bar-chart programme, which extends over the entire life cycle of the project or contract.

(2) Refine the base-case cash flow to take account of delays between incurring a commitment and paying or receiving the money.
(3) Calculate the resulting cost and benefit together with the investment required.
(4) Consider the implications of risk and uncertainty.
(5) If necessary, examine the implications of inflation.

In particular, inflation should always be considered separately when all other aspects of the investment are understood. Additional assumptions

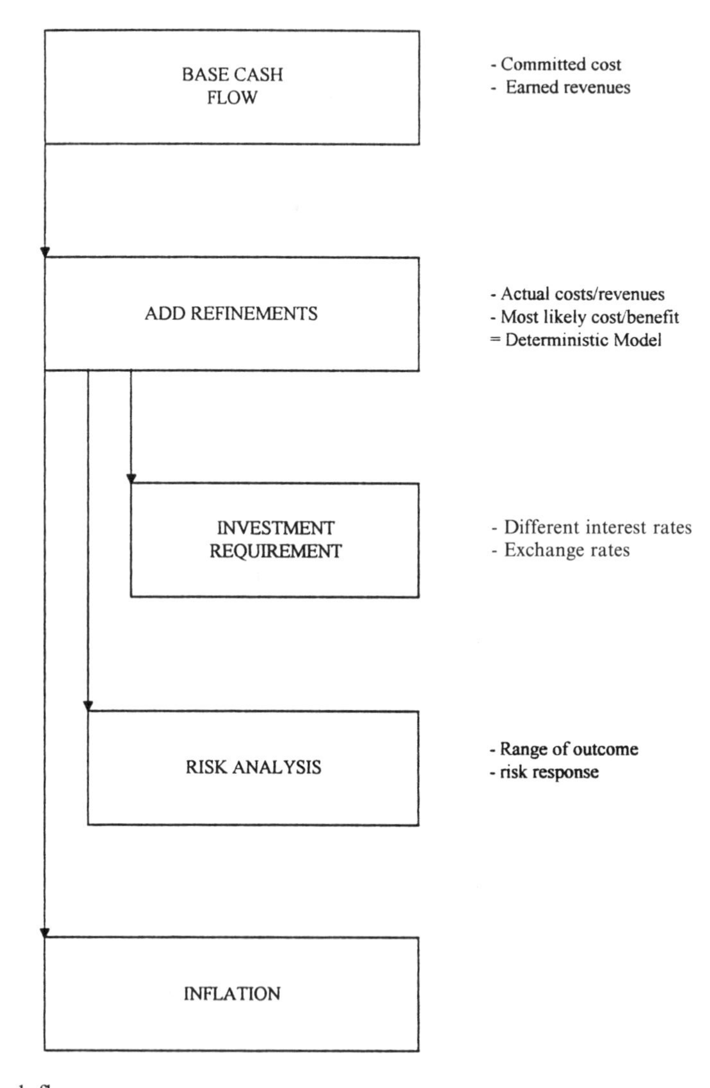

Figure 7.1 Modelling cash flows.

concerning future rates of inflation then have to be made, which increase the uncertainty. It is easy to be confused or misled if inflation is introduced too early in the financial analysis.

7.2 Categories of charge

There are three basic types of cost or revenue that make up the cash flow:

(1) fixed charges incurred at a point in time;
(2) quantity-proportional charges related to quantity of work completed, output or deliveries of materials (only material costs and product revenues are strictly in this category);
(3) time-related charges – predominantly the cost of resources. These are increasingly important, particularly in resource-dominated jobs. People and resources are normally employed or hired by the week or month.

Any flow of money can be defined in one of these categories, and the realism of the cash flow prediction will depend greatly on the correct definitions of charges. If in doubt, ask yourself how the bill will be paid.

7.3 Compiling the base-case cash flow

The most likely estimates of cost and revenue are added to each activity in a realistic programme. Initially keep the programme simple by splitting the work into a small number of major activities.

Set the base, or reference, date for the estimate, normally the date of the first flow of money, and divide the contract or project life into time periods appropriate to the accuracy required. Economists may compute annual cash flows in the early stages of project evaluation but it is normal to consider monthly cash flows when appraising or managing the project, while it is advisable to calculate weekly cash flows for short-duration contracts.

The base cash flow is compiled using the costs and exchange rates as they were at the base date. Distinguish between fixed, quantity-proportional and time-related charges, and when necessary convert all cash flows to one currency. Assume zero inflation.

The cash flow patterns should be sketched on each activity in the programme, prior to computing the period cash flows and constructing the cumulative curves. These should be completed before proceeding to add the refinements introduced below. The process is illustrated in the following example, and Figures 7.2 and 7.3.

Example: Contract for the supply of a machine

The contract for the manufacture and installation of a machine has been defined in Figure 7.2 as five activities on a bar chart, i.e. design, fabrication, installation and commissioning of the machine plus construction of the foundation. The 26 week programme is known to be tight, and the contract includes modest liquidated damages.

The costs associated with the completion of each activity are calculated and the resulting cash flow patterns sketched onto the bar chart utilizing the appropriate cost categories.

In this example:

- It is assumed that a design team is allocated to this job for six weeks. They generate a time-related charge of £-/week, which is shown to accumulate linearly. If the activity takes longer than predicted, then the cost will extend (and subsequent activities may be delayed).
- Fabrication of the unit incurs several different costs: a fixed cost for setting up the job, time-related charges for labour and the use of the facility, and material charges as they are scheduled to be delivered, all as sketched against this activity on the bar chart. The reader will recognize that by categorizing the costs the consequence of a delay or of an increase in material costs can quickly be deduced.
- Similar cost patterns are estimated for the other activities.
- General costs associated with several activities have been isolated into 'indirect' cost centres or 'hammocks'. It is here assumed that the same crane will be retained on site for construction of the foundation and installation of the unit, also that overheads are incurred as a weekly charge while work is proceeding on site. These time-related hammock costs are an important mechanism in the development of this simple 'time-and-money' model. They should be viewed as elastic, and they are of course sensitive to changes in programme. All general resources should be separated as hammocks in any model.

The total weekly costs are then accumulated to give the cumulative cost curve, as shown in Figure 7.3. This is the best prediction of the progressive commitment of expenditure on this contract.

The pattern of income is then added to that diagram as shown by the broken line. In this case payments are staged – on completion of design, fabrication and installation, with the largest payment at the end of the

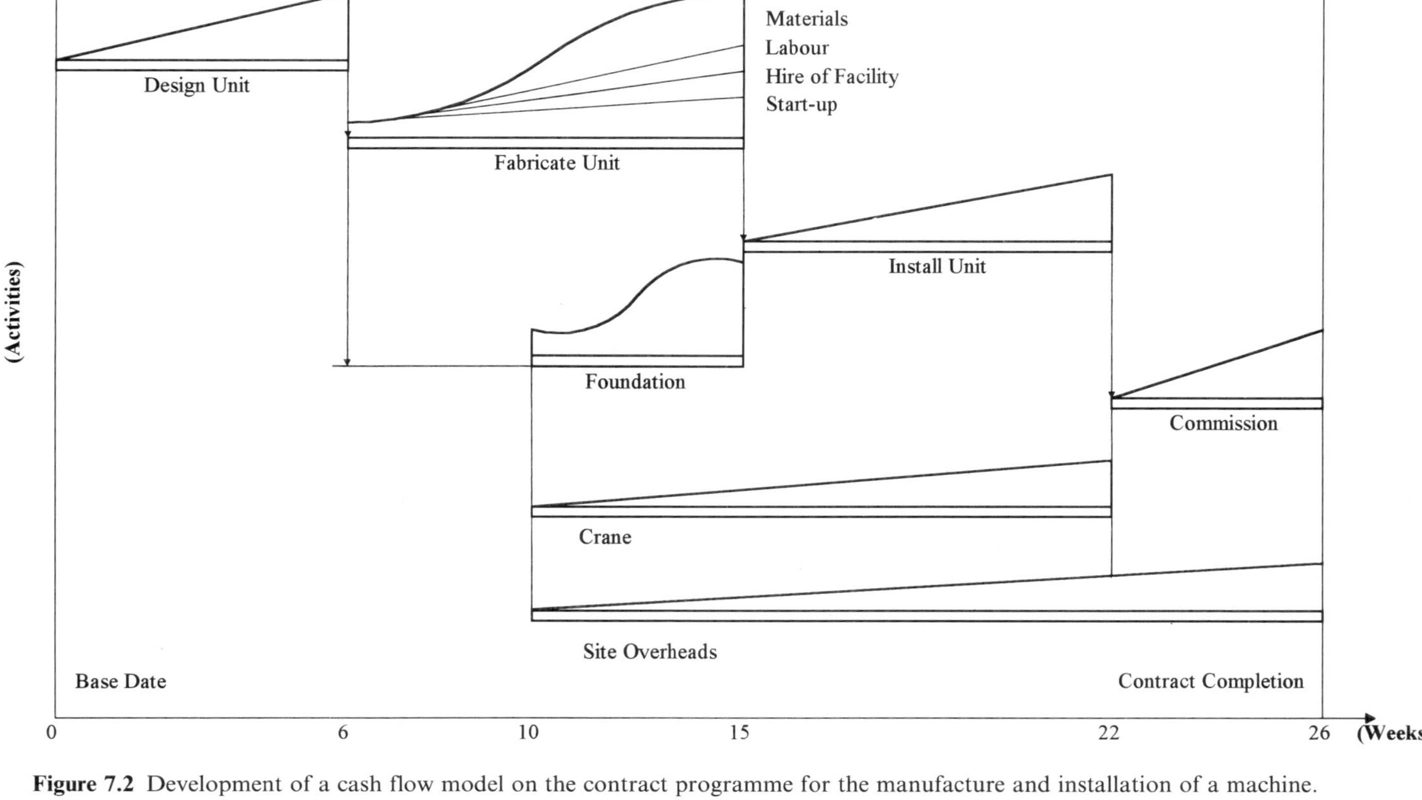

Figure 7.2 Development of a cash flow model on the contract programme for the manufacture and installation of a machine.

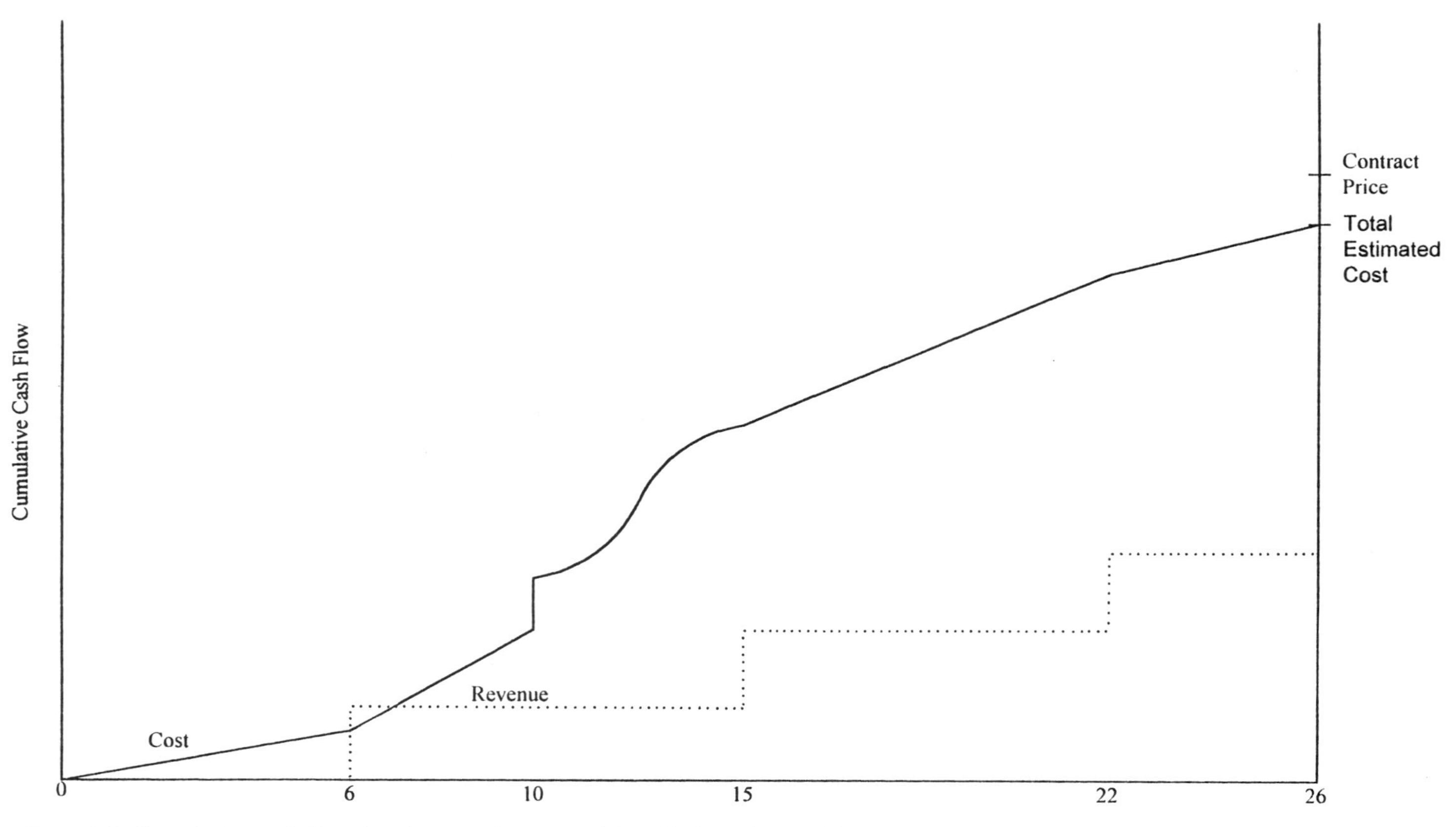

Figure 7.3 Cumulative cash flow for the manufacture and installation of a machine.

contract. This revenue curve predicts how money will be earned, and the area between the cost and revenue curves is an indication of the investment required from the contractor.

Even in this very simple form the base cash flow model offers a useful management tool, as the implication of any change is easy to visualize. In this example, adherence to the schedule is of great importance, as any delay in completion will rapidly increase interest payments on the contractor's investment, and liquidated damages may also be incurred.

Contract cash flow

It is easy to see from Figures 7.2 and 7.3 that the investment would be greatly affected by any change in the payment characteristics or any circumstances that delayed either expenditure or receipts. The contractor's cash flow can therefore be said to be 'sensitive' to such changes. Remember that cash flow in a contract that is of relatively short duration is sensitive to:

- the type of contract, for example admeasurement or cost-reimbursable;
- the contractual payment characteristics imposed by that contract, including mobilization fees and retentions;
- the delay between incurring costs and paying the bills;
- disruption of the work plan;
- the speed and extent of reimbursement of variations and claims.

Inflation is only likely to be significant if the contract duration exceeds 12 months and the underlying rate of cost escalation is 5% or more.

The investment and confidence in the cash flow forecasts will also be affected by uncertainty surrounding any aspect of the contract. A full quantified risk analysis of the type applied later in this chapter to a project cash flow of long duration will rarely be necessary for a small contract, but risks must not be ignored. The implications of each variable should be considered and appropriate contingencies allocated in the programme and estimate.

Refining the contract cash flow

The simple basic model will assist decision-making, as illustrated below in section 7.4, but for accounting and investment predictions it will be necessary to adjust the base cash flow to take into account supplier credit and the lag in payment of earned revenue. The actual cash flow will be later than shown in our basic model. In most cases the delay in payment of revenue (e.g. in accordance with the terms of the contract 'payment will be made within x weeks of submission of the certificate') will be greater than the lag in expenditure, as most workers expect to be paid weekly!

The refined cash flow predictions linked to a specified rate of interest will generate figures of the investment required to complete the contract. This becomes an element of the contractor's cost and must be included when determining the contract price.

7.4 Cost control

Cost control is one element of the overall process of management of investment in the project or contract. It relies on careful planning of the allocation and commitment of resources linked to an appropriate policy for the procurement of materials. The overall management process must take into account the agreed objectives and the commercial requirements of the job.

Contractors' cost control traditionally consists of recording costs in order to prompt corrective action where there are variations from a budget – which is itself an earlier forecast of cost. If the primary objective of control is economy then management effort must be directed to profit-making operations being carried out at unnecessarily high cost as well as to the loss-making operations. 'Dynamic control involves judgement and action to avoid variances. It demands action in time to anticipate and, if possible, avoid extra costs. This is the concern of those responsible for any section of a contract. If applied rigorously it will ensure that nobody commits resources and spends money without first considering the consequences' (Kelsey and Evans, 1989).

The essential requirements of a cost control system are:

- that it is simple and readily understood;
- that it does not consume excessive resource. It must also encourage honesty and commitment, i.e. not be seen as a chore, and must obviously lead to benefit.

Effective cost control depends on many factors, which can be identified by reference to Figures 7.2 and 7.3.

(1) *The estimate* must be realistic and so constructed that future expenditure can be monitored against it. In many cases it will be

converted into a control document, which is the intended purchase format.

(2) *Contingencies* for areas of uncertainty or potential difficulty must be allowed in the estimate and authority for their use clearly defined.

(3) *Cost forecasting.* A cost still to be incurred is of paramount importance in the financial control of the contract. As the contract progresses the original forecasts contained in the estimate turn into actual commitment and there will inevitably be variances between the bid. These variances must be taken into account when forecasting future expenditure on the remainder of the contract.

(4) *Variances* fall into three broad categories
- design and estimating variances, which are usually the responsibility of the contractor, and lead to profit erosion. Feedback for correction of the estimating database is essential.
- unforeseen costs, which may or may not be retrievable from the promoter.
- it is important that appropriate records of them are maintained.
- underspending/overspending.

(5) Agree the price of variations required by the promoter before undertaking the work. This will provide incentive for efficient performance by the contractor and also encourage prompt payment by the promoter.

(6) Orders must be placed in accordance with the contract programme, and every operation must be expedited to prevent a supplier or subcontractor causing a delay.

(7) Estimating design input is always difficult, especially when the work is innovative. The manager must also discourage 'designing for perfection', when this exceeds the required specification, and 'designing the easiest thing that can be designed rather than the easiest thing that can be built'.

(8) *Investment.* Financing charges are a significant element of contract price. They are sensitive to all forms of delay – prolonged use of resources, delay in completion of work, late submission of invoices and slackness in requests for payment for variations and claims. Remember the importance of achieving the programmed inflow of money.

(9) *Monitoring and reporting* of any factor such as output or cost by the measurement of actual achievement and comparison with forecast achievement does not in itself constitute control. It is however an essential part of the control cycle, the basis of reports

of progress to various levels of management, and provides the information needed for updating of the plan. The monitoring system should focus attention on future work to be completed. Maximum effort should be devoted to updating the plan.

(10) *Planning.* Speed of feedback of data to the planner is important and simple, even approximate, criteria should be devised and utilized to quantify progress. Here we are primarily concerned with:

- critical activities – as delay here will directly affect the completion date for the whole job;
- activities of long duration – for it may here be possible to improve performances. Little can be done in this respect on any activity of less than four weeks' duration;
- activities that consume expensive resources.

Allocate responsibility for reporting on specific activities to named individuals. Responsibility for developing contingency plans for known potential problem areas should be allocated to individuals in the same way. Not only does this procedure ensure that thought is given to the problem, it also ensures that more individuals are actively involved in planning the work.

The information required by the manager about key activities is:

- Planned start date
- Planned finish date
- Actual start date
- Quantity of work completed to date
- New estimate of quantity of work remaining
- Predicted completion date.

If the activity is likely to overrun, what action could be taken or what additional resource is needed to achieve the planned completion date?

Most contracts and projects are completed against a demanding timescale, and the old saying 'control time and cost will take care of itself' is very true. Most resource costs are time related and consequently any extension of programme will lead to increased cost.

Opportunities should be sought to introduce float into the programme. In the contract example given above it may be possible to introduce a small overlap between design and fabrication, to accelerate fabrication and/or to construct the foundation earlier. The cost of making those provisions to reduce the risk of late completion can easily be determined from the cash flow diagrams.

7.5 Project cash flow

There is some similarity between the contractor's balance of account for the above contract and that for the promoter developing a new facility or project, which is shown by the general curve in Chapter 2. In each case, a period of investment is followed by a period of surplus in the account. The time-scale of the project is, however, much longer and, consequently, this investment is unlikely to be sensitive to delays of a few weeks in payments. The sensitive factors are here likely to be:

- market forecasts of the quantity and price of the product;
- staging the developing, i.e. the project is constructed in either one or a number of separate stages;
- operational efficiency and reliability;
- delay in commissioning;
- inflation.

In this case adjustments to the base cash flow for the lag in payments are more likely to be introduced in the operational phase of the project, perhaps to allow for the delay between producing the product and selling it. Emphasis must here be given to the analysis of risk.

7.6 Profitability indicators

Although it is sometimes possible to compare different investments by reference to their cash flow diagrams, it is advisable to employ and tabulate several of the various profitability indicators that quantify the investment when choosing between alternatives.

Profit – the arithmetical difference between total payments and total receipts – is an obvious case. Maximum profit could be the criterion selected for the choice of one of several alternative plans, all of which satisfy the time and resource requirements. The profit figure does not, however, give any measure of the investment required or of the effect of time on the flow of money that constitutes that investment.

Similarly, maximum capital lock-up and payback period which quantify, respectively, the maximum demand for capital and the time taken from the start of the contract or project for the investor's account to move into credit, could be used. By reference to Figure 7.4, we see that jointly these three figures define the key ordinates on the cumulative net cash-flow curve; none of them, however, takes into account the time value of money. By this we mean that it is better to receive £1 today

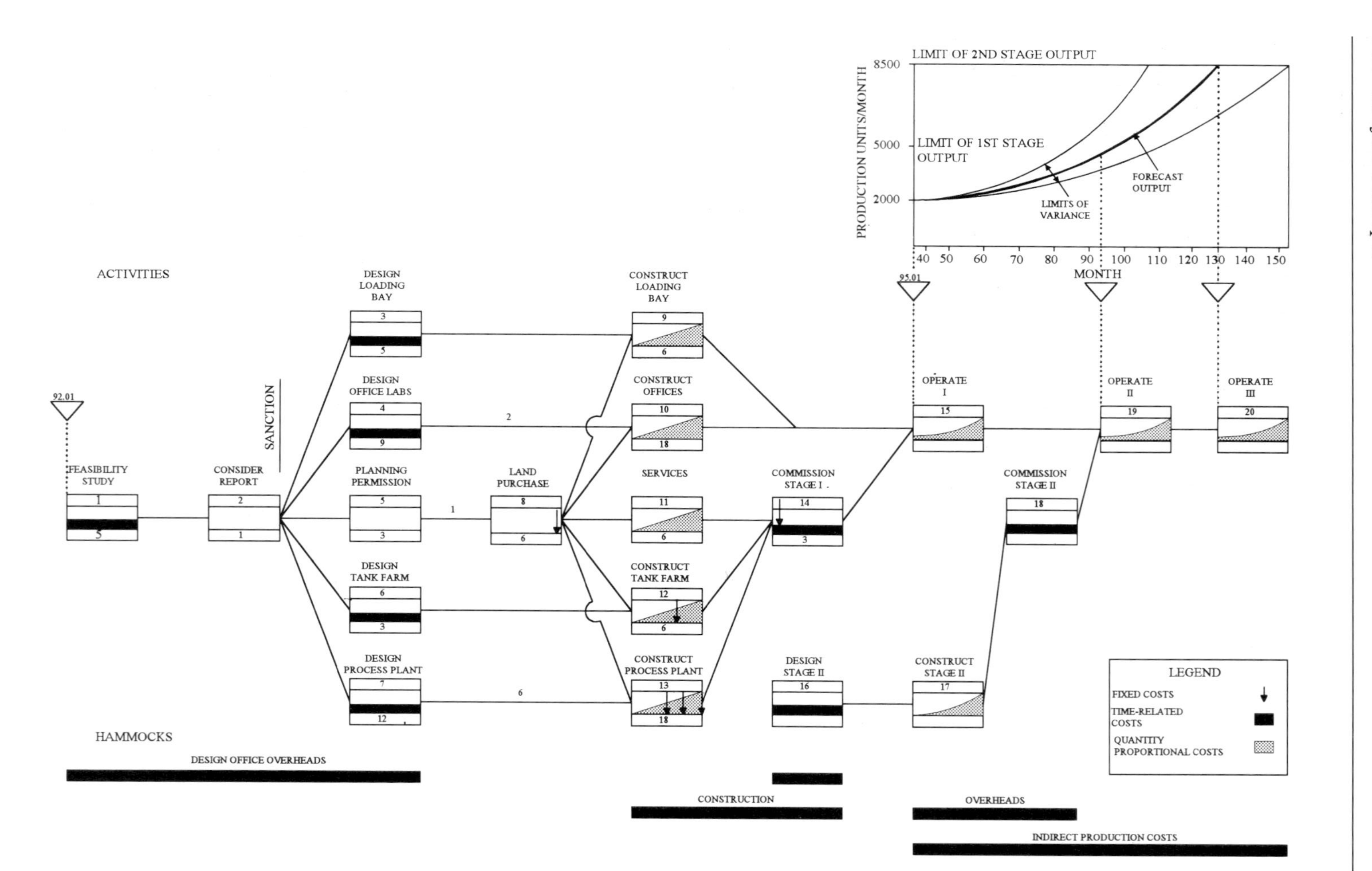

Figure 7.4 Flow chart for new industrial plant.

rather than in 12 months' time, for we will have had the use of the money in the intervening period.

The process we use to introduce a measure of the time value of money into our calculations is called *discounting*, which is briefly described in the appendix to this chapter. The individual period cash flows incurred over the duration of the contract or project are converted to their equivalent values at a single point in time – normally the start of the investment.

It is probable that either net present value or internal rate of return would be used in conjunction with several of the other criteria for the selection of a project or to identify a preferred contract cash flow. For example, a positive net present value or some specified minimum value of internal rate of return may be required in addition to a minimum level of profit. A short payback period is also desirable and is given prominence in the selection of many commercial projects. The use and tabulation of several profitability indicators is strongly advised whenever comparing investments.

The author has also found the total investment, the area below the horizontal base line expressed in £-years, to be useful during project appraisal. The total investment in new industrial plant is the area of the triangle = approximately 37.5 million £–years.

The same criteria are also a convenient way of evaluating change in either contract or project. In both cases, the investor is greatly concerned with the effect of change on his investment, and all the factors to which cash flow was said to be sensitive in Section 7.4 above may be assessed in this way.

7.7 New industrial plant

The design and construction of this hypothetical manufacturing plant have been divided into 14 activities, and the resulting flow chart (Figure 7.4) has been extended to include the subsequent operation of the plant. The time-scale therefore spans the entire life of the project envisaged from conception to the end of operation. The activities are linked in normal precedence form and overlaps defined. The required commissioning date of 1995.01 (the first month of 1995) is a major constraint on the development of this project. The general cost patterns associated with individual activities are indicated on the diagram and indirect costs (which are entirely time related) are defined as hammocks below. In the operational phase overhead costs for staff, maintenance and the general running of the plant will be of this type (the indirect production costs)

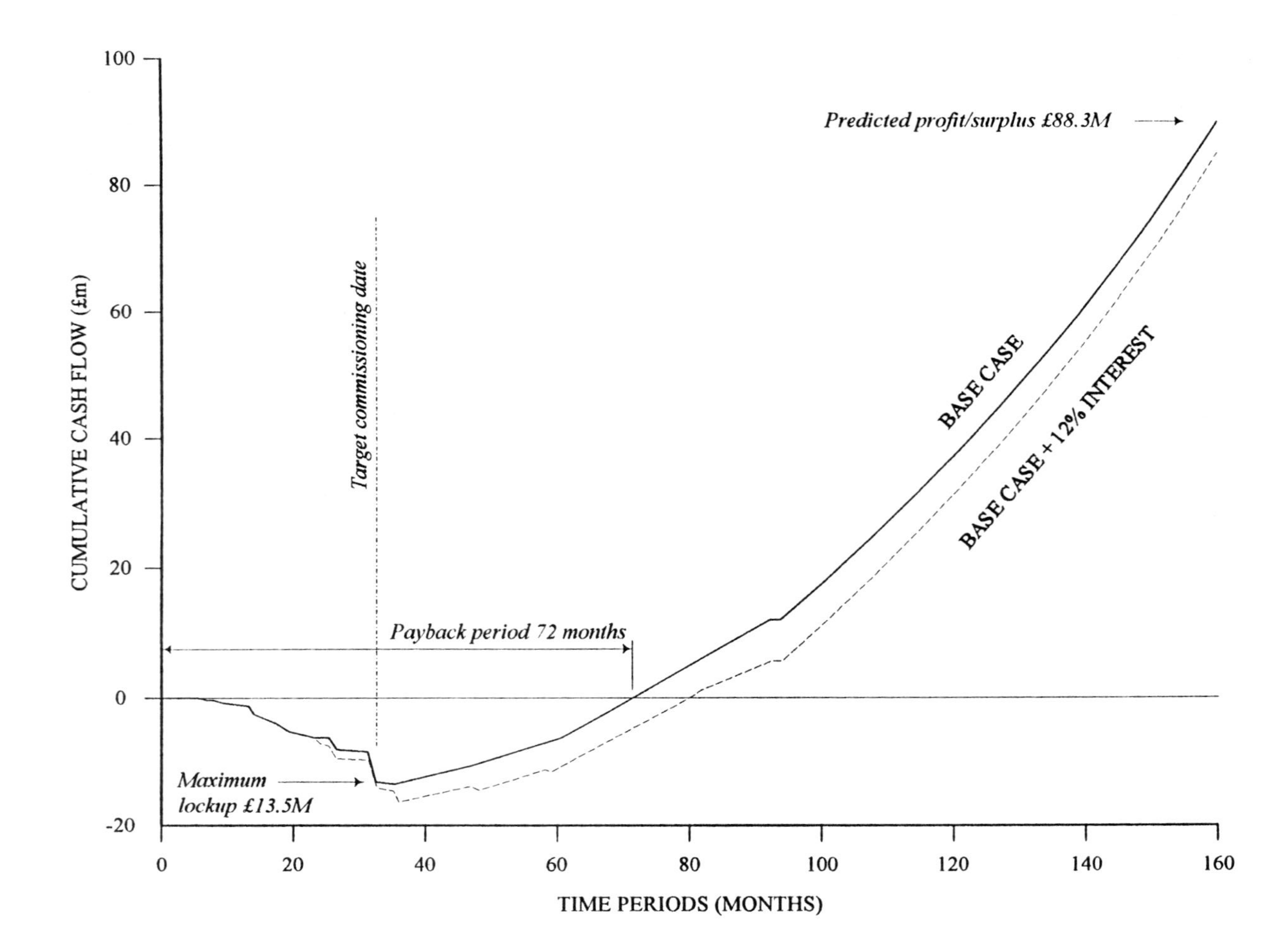

Figure 7.5 Cumulative cash flow predicted.

but there will be a considerable element of direct cost related to the quantity of production. Revenue will of course be related to the number of units produced and sold.

The production diagram shows that the maximum capacity of the plant erected in the first stage of the development is expected to be 5000 units per month and that the initial demand for the product is predicted to be 2000 units. The target curve of forecast output is based on a growth in demand of 20% p.a., but attention is drawn to the uncertainty surrounding this figure by the outer curves showing the likely rate of increase to vary between 15 and 25% p.a. The second stage of the development, which increases the capacity of the plant to 8500 units/month, will be constructed at a later date if justified by the proven demand for the product. Should all these predictions prove optimistic, the second stage may not be required at all!

The model is required to simulate the implications of alternative patterns of project development and to predict the consequences of risk in terms of the effect on the investment. It is designed to develop interactive time and money relationships and to give a realistic response in terms of cash flow to change of all the variables considered. This is initially a 'paper-and-pencil' qualitative exercise involving all persons responsible for providing estimates or predictions. Thereafter the data can be assembled using a computer program, such as CASPAR, in order to assist simulation of change.

The best estimate

A range of financial criteria, including both cash and discounted values, should be used to quantify the investment.

A typical selection are adopted for this study in which the 'best' single figure prediction is a cash surplus of £88.3m, a payback period of 6 years, investment of 37.45 £m-years and an IRR of 36.8%. This most likely forecast of the cost–benefit of investment in this project, generated from the refined base case model, is shown in the cumulative cash flow diagram (Figure 7.5).

Interest payments have been excluded from the basic simulation but their effect is shown by the dotted line; the payback period is increased by 9 months and the IRR reduced to 31.3%. This return will, however, only be achieved if all predictions and estimates over the 13-year period of appraisal, implementation and operation of the project are precisely fulfilled!

Consideration of risk and uncertainty

It is likely that there will be uncertainty about many of the predictions incorporated in the base model, and these will now be investigated. At this stage it is possible to obtain a feel for the implication of these uncertainties simply by substituting different values for variables in the original model, recalculating the cost and benefit and sketching the new cash flow curve. This repetitive exercise can be used to predict the effect of delay in commissioning or a change in market forecasts even if a full risk analysis using a computer is not attempted.

The risk arising from the market or demand for the service or product may be reduced by staging the development of the project. Implementation of the second stage of new industrial plant will depend on the actual production figures achieved following the commissioning of stage 1. Attention should also be paid to risk and expenditure at the end of the project life. The cost of decommissioning may be significant in cash terms but can easily be ignored if only discounted figures are studied.

In the case study of new industrial plant the CASPAR program has been used to expedite the analysis and to study the consequences of change in a selection of variables. Their relative significance in terms of effect on the viability of the project is well illustrated in the sensitivity diagram (Figure 7.6). The most sensitive factors in this project all relate to the operational phase, and the forecast of the market for the product and the selling price are particularly significant. If the selling price of the product were to fall by 20% (say from £100 to £80 per unit) the predicted profit would drop from £88.3m to around £38m! Timely completion is also of crucial importance. This is frequently the case. The viability of projects entering a competitive consumer market is greatly dependent on realistic market forecasts and on early/timely commissioning of the facility.

Sensitivity analysis is a useful means of assessing the risk from specific elements of the project and normally precedes probabilistic analysis.

Referring to the sensitivity diagram for new industrial plant:

- the relative importance of each variable is immediately apparent;
- individual risks are displayed on one diagram, facilitating the identification of interrelated risks;
- judgement can be made of the likely range and occurrence of individual risks.

(The sensitivity or 'spider' diagram should be plotted in small increments over the likely range of variation of each factor. In many cases they will

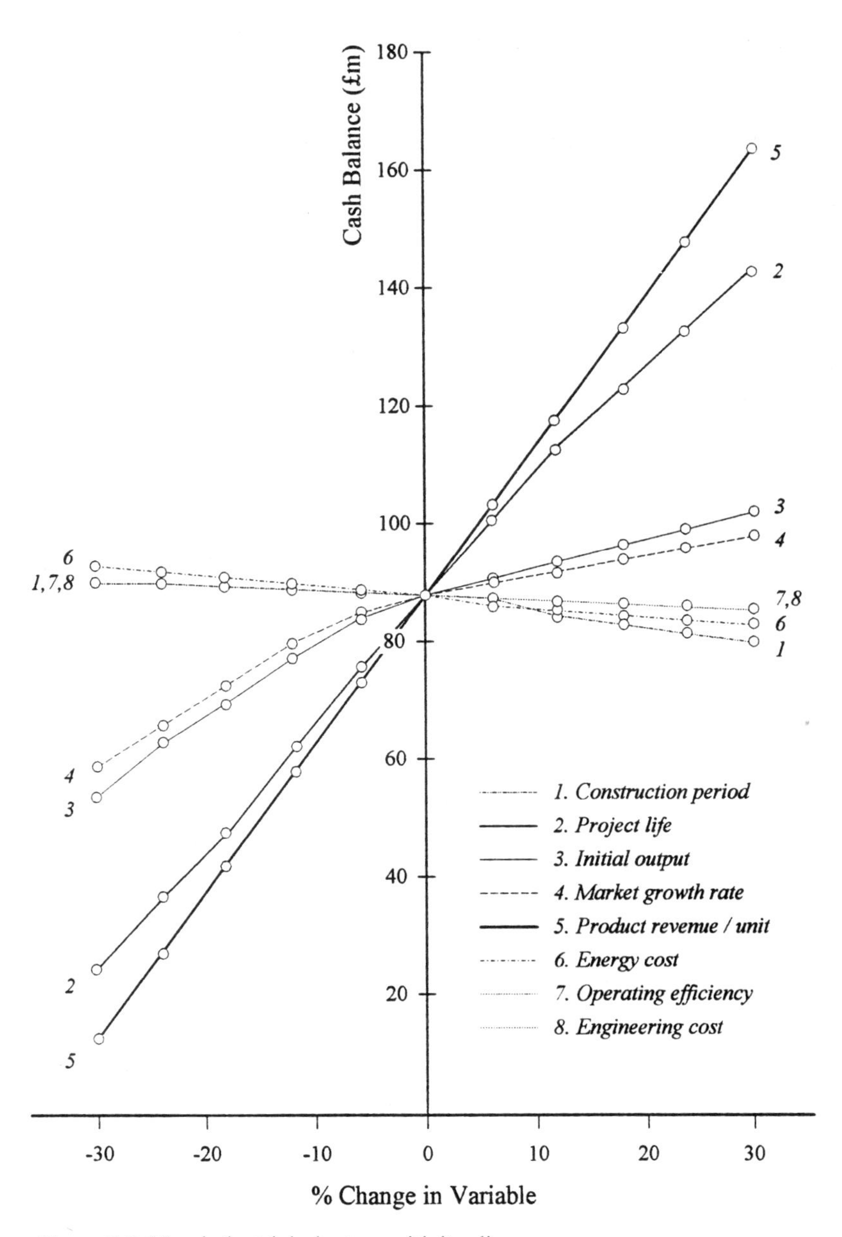

Figure 7.6 New industrial plant: sensitivity diagram.

not generate a straight-line relationship, and any significant change in gradient should be investigated. The origin of the graph, in this case £88.3m, is the profit predicted when all the assumptions made in the base case estimates are achieved.)

Already, this simple assessment of the individual risks can influence management strategy for the further development of this project. Additional market surveys may be initiated to increase confidence in the forecasts and, during the engineering phase, emphasis must be given to achievement of programme even at the expense of additional cost.

In practice it is likely that some combination of the uncertainties considered individually during sensitivity analysis will be encountered. It is therefore desirable to make some assessment of the implications of the combined risk, and this has been done by assigning probabilities of occurrence to each of the major variables and performing a probabilistic analysis. The analysis is of the Monte Carlo type whereby 1000 iterations have been simulated and the results presented as a cumulative frequency diagram in Figure 7.7.

The allocation of probabilities of occurrence to each of the variables included in the risk analysis is subjective and should involve the personnel responsible for the original single-figure estimates. The range and pattern specified for each variable indicates the degree and nature of uncertainty about the original predictions perceived by the initiator.

Interpretation and perception of risk

The results of the risk analysis as shown in Figure 7.7 should be interpreted as giving a guide to the likely out-turn of the project. The original single-figure prediction of 36.8% is seen to be optimistic as there is a 60% probability that the IRR will be less than this value. It is suggested that a range estimate should be substituted. In this example it would be more meaningful to quote a mean IRR of 32% and a 15/85% range of 10–55%: i.e. there is an 85% probability of achieving an IRR of less than 55% and only a 15% probability of the IRR falling below 10%.

It must also be accepted that not all risks are included, and the exclusions should be acknowledged. For instance, only a small range of operating efficiency has been considered appropriate for this plant, but in the extreme case, if the plant failed to produce an acceptable product, even for only a short period, the market could be irretrievably affected.

Study of the relative sensitivity of variables on a common scale, as in Figure 7.6, is valuable and, as mentioned above, will aid definition of project management policy. It is however also necessary to consider, 'think through' and understand the implications of the individual risks and uncertainties.

The analysis should be revised at regular intervals during project implementation. Ideally, up-to-date forecasts should be provided by those responsible for the original estimate. Reappraisal will also ensure that project management continue to work for realistic targets.

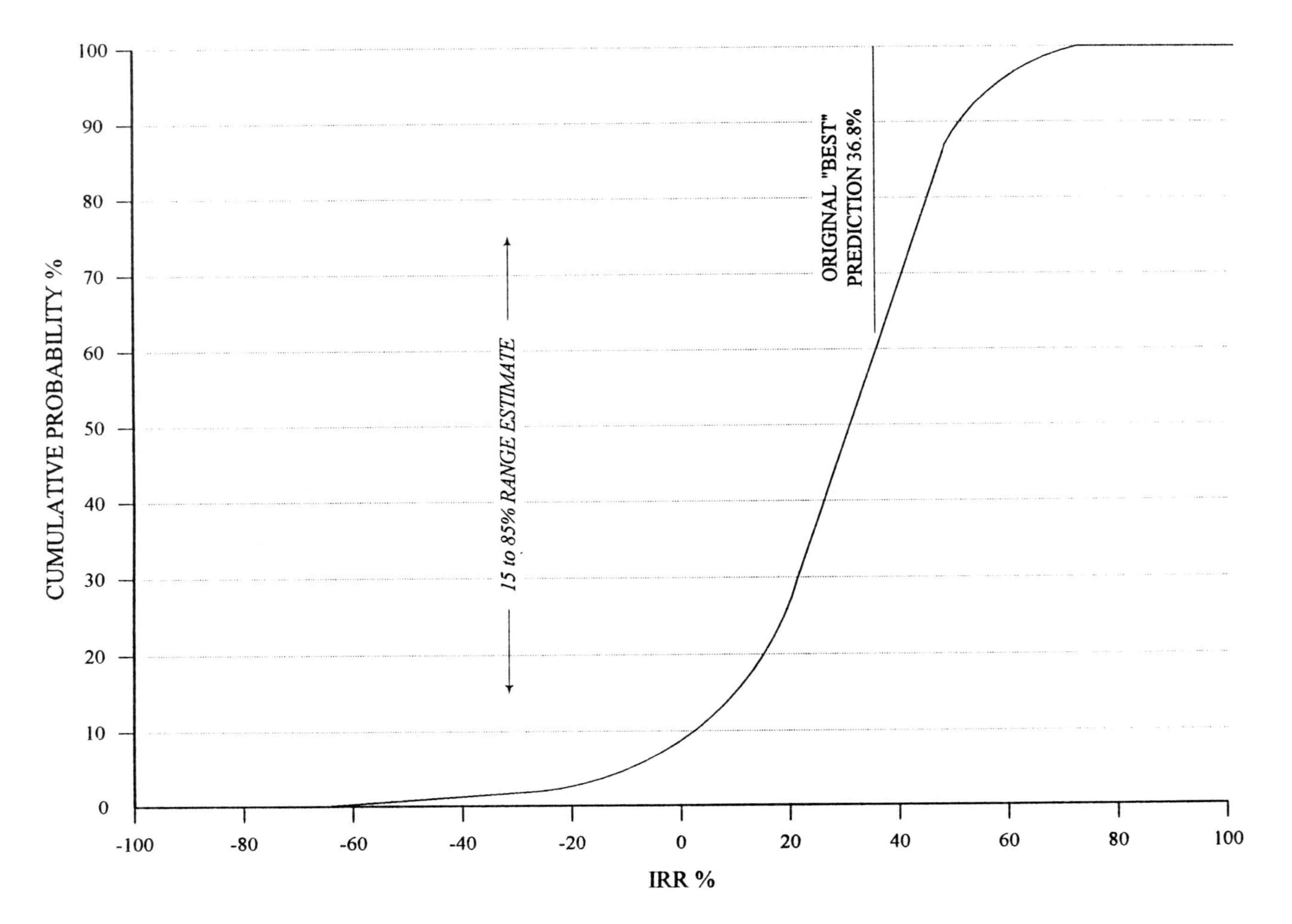

Figure 7.7 New industrial plant: cumulative frequency diagram.

7.8 Inflation

Inflation is the decrease in the purchasing power of money arising from escalation of costs and prices. The effect is compound, cumulative, and is consequently extremely time-sensitive. When base-case cash flows are escalated to predict 'money of the day figures' the increases can appear dramatic: they can also be misleading!

Figure 7.8 illustrates this effect when the base-case costs and prices for new industrial plant are all increased at an inflation rate of 10% p.a. In 'money of the day' terms both the demand for capital and the predicted profit increase significantly, the former from £13.5m to £16.2m and the latter from £88m to £240m. The apparent improvement in benefit is, however, a myth as the purchasing power of money will decrease over the life of the project.

The difficulty of predicting future rates of inflation is obvious when one considers the technical, political and economic changes of the last decade – all of which affect commodity values, exchange rates and market confidence. Different elements of cost will also escalate at different rates: for example, changes in the cost of labour or fuel may not be reflected in the cost of imported materials or construction plant. Published national 'rates of inflation' are normally linked to a collection of domestic prices to indicate movement in 'the cost of living' and may be very different from the costs of raw materials used by your project.

Our recommendation is that:

- For projects of more than 10 years' overall duration and particularly where the cash flow analysis is performed primarily for ranking purposes, escalated forecasts should be given relatively little weight and the investment decision based on the non-escalated figures.
- In all cases the maximum escalated demand for capital should be determined.

Fortunately for the project manager he or she is unlikely to be required to generate 'money of the day' figures when estimating project costs and benefits. This is more often the responsibility of economists or accountants, who may simulate the implications of a range of inflation rates. Tax and/or royalty payments levied on the profits generated by your project will, of course, be calculated at 'money of the day' values.

In engineering contracts the risk arising from variations in exchange rates should be clearly allocated to one of the parties. For contracts exceeding 12 months in duration it is normal to compensate the contractor by the application of a contract price adjustment formula linked

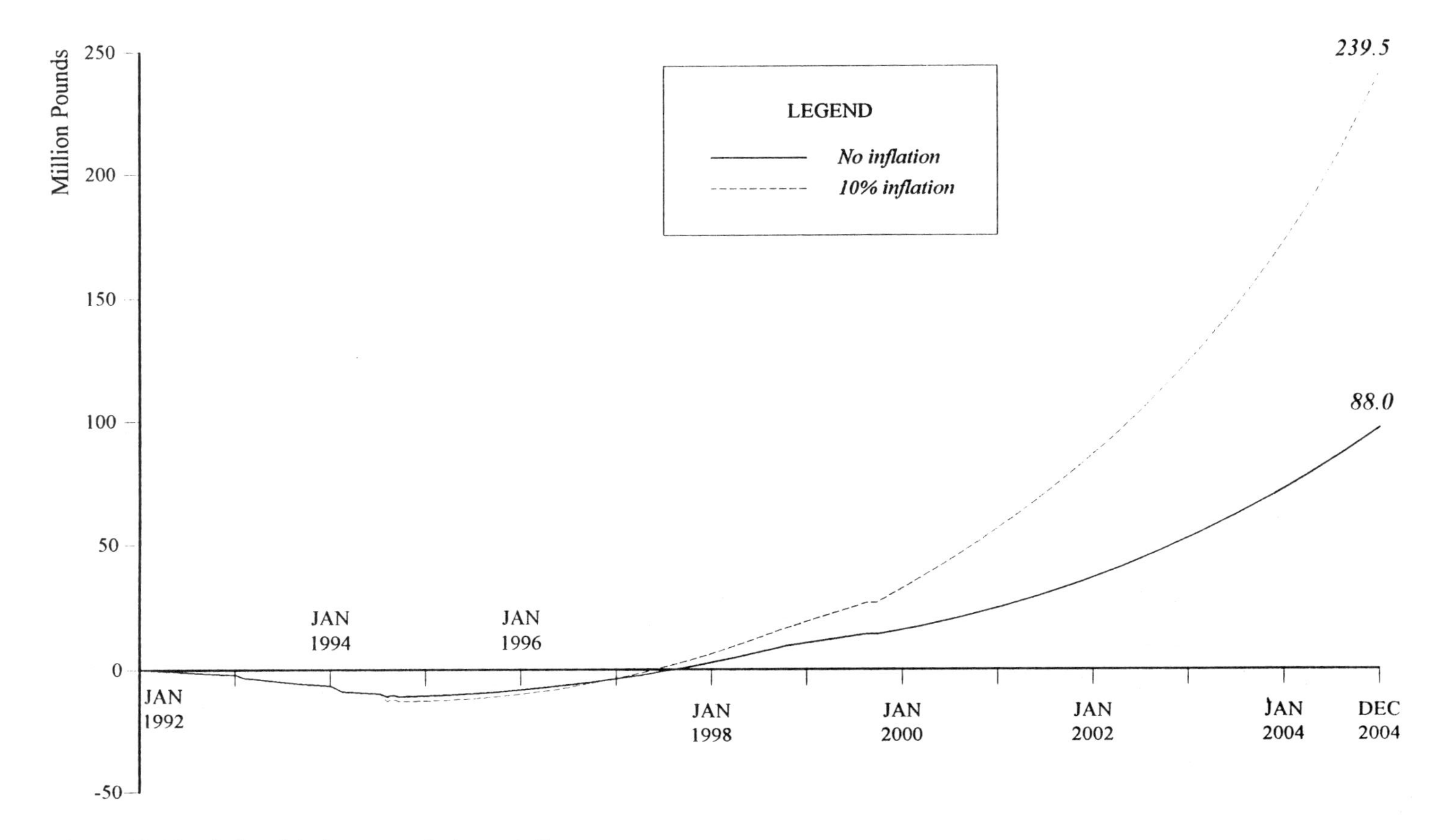

Figure 7.8 New industrial plant: cumulative cash flows.

to published national indices. The UK 'Baxter' formula incorporates 12 cost categories common to civil engineering works and is a good example of an equitable and simple method of compensation.

7.9 Risk management

The simple time and money models described and developed in this chapter are valuable when defining and evaluating the uncertainties associated with the estimates and predictions for any project or contract. The reader is now referred back to Chapter 2 to reconsider the process of risk management.

All engineering projects and most contracts are subject to uncertainties, which must be investigated if realistic estimates and responses to risk are to be generated. The examples given in this chapter have illustrated that uncertainties such as resource output can give benefit as well as loss, and that many risks in the implementation phase result in delay. It should also be noted that the earlier in the project or contract life action is taken to reduce risk, the more effective it is likely to be.

It is relevant to emphasize the following selected conclusions drawn from recent research (Perry and Thompson, 1992).

- All too often risk is either ignored or dealt with in an arbitrary way: simply adding a 10% 'contingency' onto the estimated cost of a project is typical. This is virtually certain to be inadequate.
- The greatest uncertainty is in the earliest stages of a project, which is also when decisions of greatest impact are made. Risk must be assessed and allowed for at this stage.
- Risks change during most projects. Risk management should therefore be a continuing activity throughout the life of a project.
- Much can be learned about the implications and management of project risk without extensive numerical analysis. Risk analysis is essentially a brain-storming process of compiling realistic forecasts and answers to 'what happens if?' questions.
- The quantitative assessment of risk requires analysis of the likely extent and interaction of variable factors. The analysis should be carried out by those trained to do so jointly with project planners and cost estimators. The need for judgement should not be used as an excuse for failing to give adequate consideration to project or contract risk.
- Delay in completion can be the greatest cause of extra cost and of loss of financial return and other benefits from a project. The first

estimate of cost and benefits should be based on a realistic programme for a project. On this basis the potential effects of delays can be predicted realistically.

- The overriding conclusion drawn from the research is that all parties involved in construction projects and contracts would benefit greatly from reduction in uncertainty prior to financial commitment. Money spent early buys more than money spent late. Willingness to invest in anticipating risk is a test of a client's wish for a successful project.

Further reading

EDC for Civil Engineering (1973) *Price Adjustment Formulae for Civil Engineering Works*, HMSO, London.

Gaisford, R. (1986) Project management in the North Sea. *International Journal of Project Management*, **4** (1), 5–12.

Kelsey, K.J. and Evans, T.J. (1989) Financial control in mechanical contracting. In: *Control of Engineering Projects* (ed. S.H. Wearne), Ch. 5, pp. 99–124. Thomas Telford, London.

Merrett, A.I. and Sykes, A. (1973) *The Finance and Analysis of Capital Projects*, Longman, London.

Norris, C., Perry, J.G. and Simon, P. (1992) Project risk analysis and management. *Association of Project Managers Bulletin*, March.

Perry, J.G. and Thompson, P.A. (1977) Construction finance and cost escalation. *Proceedings of the Institution of Civil Engineers*, **62**, 623–624.

Perry, J.G. and Thompson, P.A. (1992) *Engineering Construction Risks – A Guide to Project Risk Analysis and Risk Management*, Thomas Telford, London.

Thompson, P.A. and Norris, C. (1993) The perception, analysis and management of financial risks in engineering projects. Proceedings of the *Institution of Civil Engineers*, **97**, 42–47.

Thompson, P.A. and Willmer, G. (1985) CASPAR – a program for engineering project appraisal and management. In: *Proceedings 2nd International Conference on Civil and Structural Engineering Computing*, London, vol. 1, pp. 75–81. Institution of Civil Engineers, London.

Appendix

If a stream of future payments representing an investment is discounted to the present time, the present value of the investment is calculated. When both payments and receipts are discounted, the net present value is given.

The present value p of a cash flow C_n in period n is given by:

$$p = C_n \times \frac{1}{(1+r)^n}$$

where r is the discount rate per time period (expressed as a decimal).

Care must be taken to ensure that all the alternatives to be compared are evaluated on the same basis, i.e. that the discount rate, the same base date, and the same time period or interval are used. In a non-inflationary situation, the discount rate is normally selected to represent the cost of capital to the investor: the time period may be years or months for a project and months or weeks for a contract.

It is rarely necessary to calculate the discount factor $1/(1+r)^n$, as this is tabulated and is readily available in many textbooks. Calculation of the net present value is then simply a matter of multiplying the net cash flow in the first time period by the relevant factor to give the discounted cash flow and then repeating this process for each successive time period. Receipts are designated as positive and payments as negative cash flows. The net present value is the arithmetical sum of the individual discounted values.

Calculations of present value assume a discount rate, and it is sometimes convenient to utilize the internal rate of return – which is defined as the discount rate that will produce zero net present value; i.e. net present value is a measure of the gain from a project or contract, whereas internal rate of return measures gain relative to outlay.

Discounting is an expedient technique used only to aid comparison of investments. This technique for taking into account the 'time value' of money should not be confused with provision for cost escalation, which should be considered separately.

Chapter 8
Project Organization

As observed at the start of this book, the demands for the engineering capacity and for other resources required during a project are transitory. The decision to proceed leads to an increase in resources employed, but this is followed by dispersal as the work is completed. The nature, scale and complexity of activities change during the evolution of a project; so therefore does the appropriate organization of the work, within the promoter's team and in their employment of others.

8.1 Stage-by-stage needs

An organization has to be open to the inherent need to change for the cycle of work for each of its projects, first for the phase of considering ideas, then the probabilistic stages of evaluation and design, and finally the most detailed stages of physical work. Initially expertise in markets, engineering, risks and evaluation need to be concentrated on the ideas and alternatives. Once a project is selected and sanctioned, some or all of the engineering and other resources usually have to be managed economically. To do so, design, manufacturing and implementation work for the project is likely to be done by internal departments or contractors, who aim to minimize costs by using their resources continuously over several projects. There is thus a potential conflict between two objectives:

- achieving the sequence of activities essential to each project;
- sharing resources economically amongst a variety of projects.

Projects are rarely carried out in isolation from others, because of the advantages of sharing expertise and other resources. Resources should therefore be organized to try to achieve the potential benefits of concentrating on the needs of each project and also realize the potential economy of people and organizations specializing in a stage of work for many projects.

8.2 Organization

Organization should be a means of enabling people to achieve more together than they could alone.

A standard system of organization that suits all demands has not yet been evolved. Organizations vary in their objectives, culture, work and circumstances. The system suitable for one firm is not likely to be suitable for its customers or for its suppliers. Each party should have the organization to make the best use of the resources it needs to survive and succeed in its transactions with its surrounding systems. What is needed is not permanent.

Bureaucratic rules and routines help in larger organizations. They give a sense of internal order, but for an organization to be able to overcome unpredictable problems and achieve innovation demands 'organic' flexibility, informality, and uncertainty in relationships, as observed by T. Burns and G.M. Stalker. Groups of people and individuals can then change their role-relationships as needed project by project, and as stated by D.W. Conrath the planned system can 'co-opt the informal organization' as it evolves.

8.3 Specialization

In engineering, and in other sectors, jobs have become more specialized with the continuing growth in knowledge and the skills necessary for its application. Specialization is a characteristic of organizations, departments in them, professions, crafts and individuals. The division of organizations into departments and subdivisions within them is often on the basis of specialization. Figure 8.1 shows a typical set of what are called 'functional' departments in a manufacturing firm. In this case there are two successive levels of specialization. This and the following figures illustrate various forms of specialization within organizations. The systems vary greatly and so do job titles. The common principle of division of work based upon specialization is to be seen in most manufacturing, power, gas and other utility companies, contractors, consulting offices and local and central government.

In functional organizations people employed in different departments have the expertise and information needed for a project: for instance, the

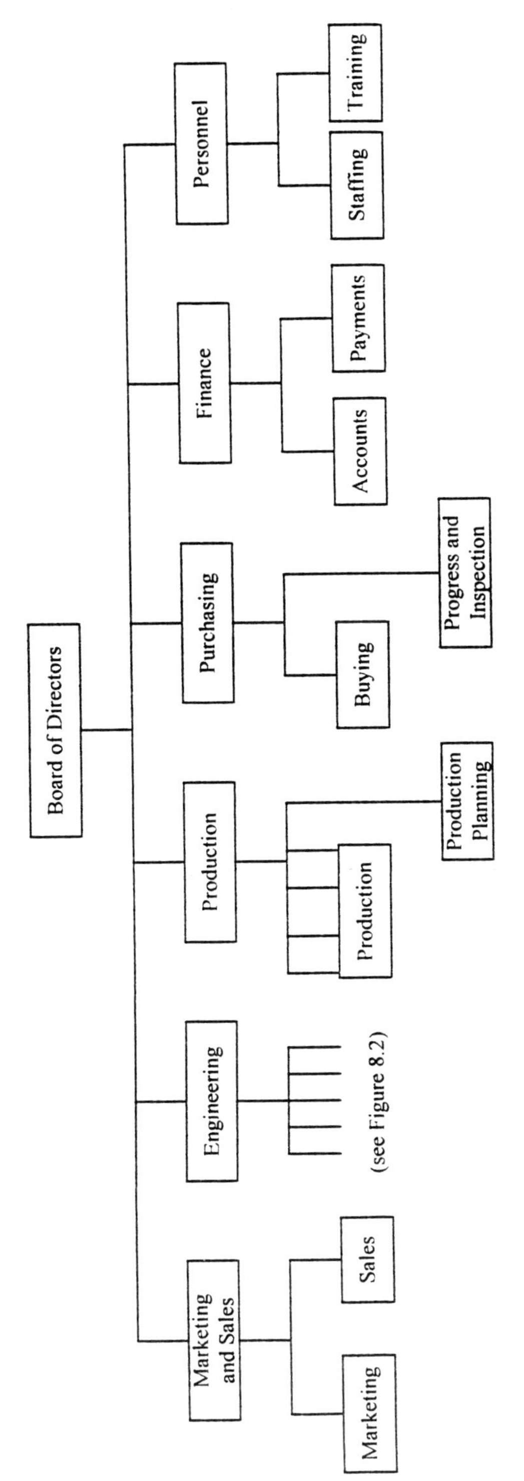

Figure 8.1 Functional organization of a firm.

marketing, engineering and production departments shown in Figure 8.1. In nearly all cases the people who design an engineering project are separated from those who undertake the subsequent activities of manufacture or implementation. Except in cases where the project is small and all of it can be done by a small team located together, the relationships between everyone contributing work to a project needs to be carefully planned and reviewed regularly.

Figure 8.2 shows a functional organization within the engineering department of a firm. The figure shows two levels for the division of work into more specialized functions. Most projects will depend upon work by several of the sections, in sequence or in parallel or both.

The division of work may be by type or size of projects, as illustrated in the role of the sections shown in Figure 8.3. In this case there is functional specialization at the next level down.

Some types of work, such as plant installation, construction and the commissioning of capital projects, usually bring together many of the people working on that stage of a project. Figure 8.4 shows an example of the organization of a civil engineering contractor's staff on a site. In this case there is division into three main sections, each employed for a distinct part of the project. The location of their work provides a logical reason for this. There is some specialization within these sections, and also in the roles of the section that provides services in site work design, safety, planning, progress and costing.

8.4 Communications

Departments in functional organizations differ in their time-scale of decisions and influence on projects. They differ in the expertise needed and the experience people obtain. They develop their own language. The people in each department may therefore differ from others in their objectives and attitudes. Success in applying their specialism can become the objective of a group or an individual when given a specialist role. Specialists who are treated as advisers rather than as members of project teams are particularly likely to be motivated to recommend safe solutions and to give priority to quality regardless of their effects on delivery times and budgets. The consequences can be problems of communication and cooperation, particularly if the departments are located away from each other or are not led by one manager.

Division of the work for a project amongst specialist groups and individuals should have the advantages of applying and developing expertise, but can therefore lead to failures of communications because

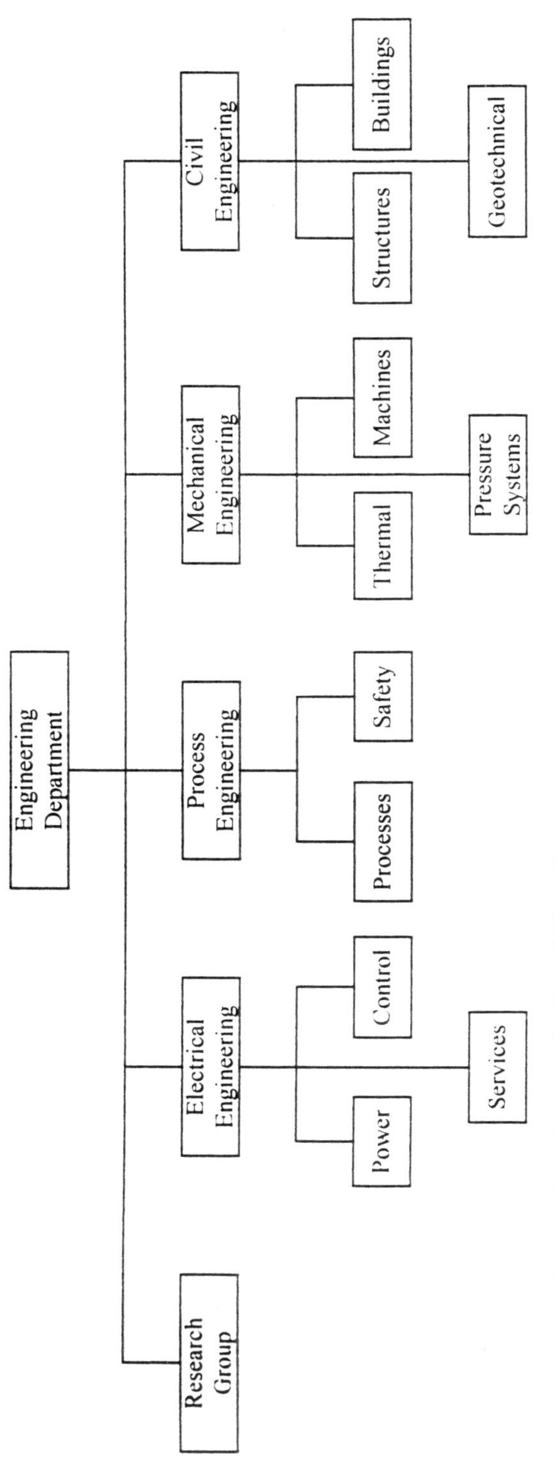

Figure 8.2 Functional engineering organization within a firm.

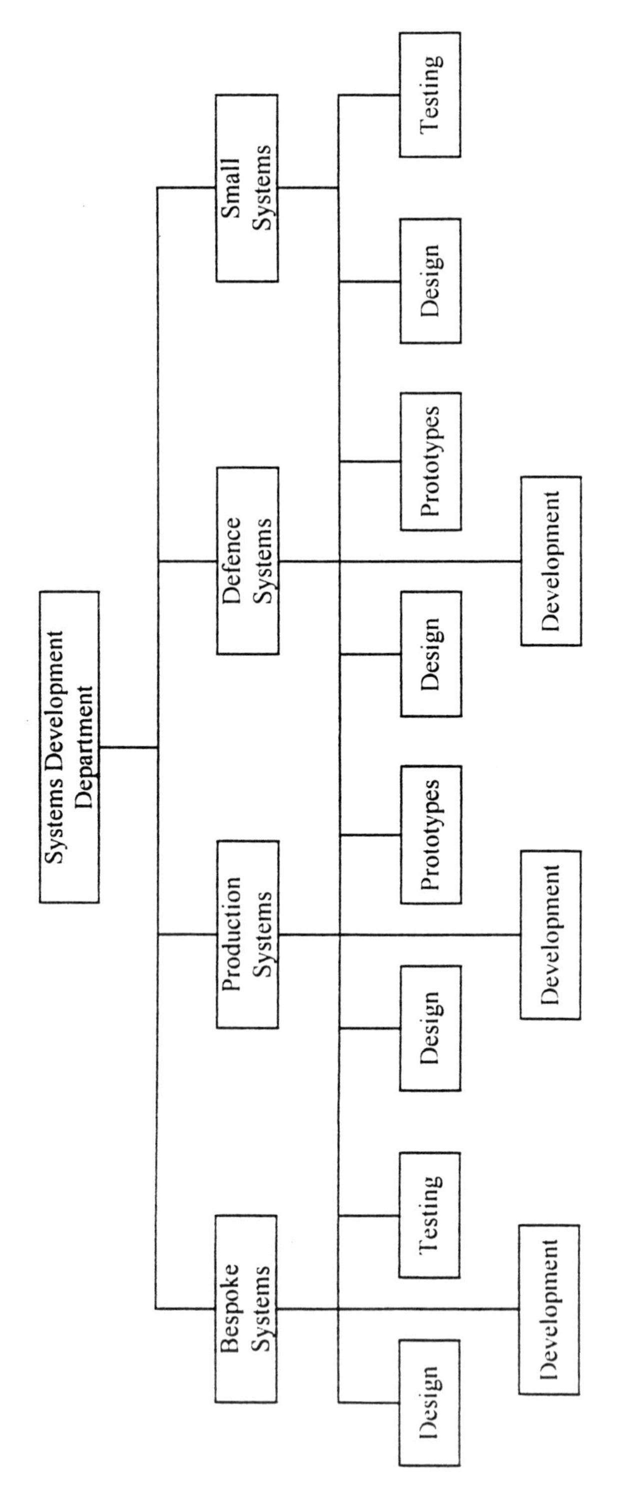

Figure 8.3 Division of work by type and size of project.

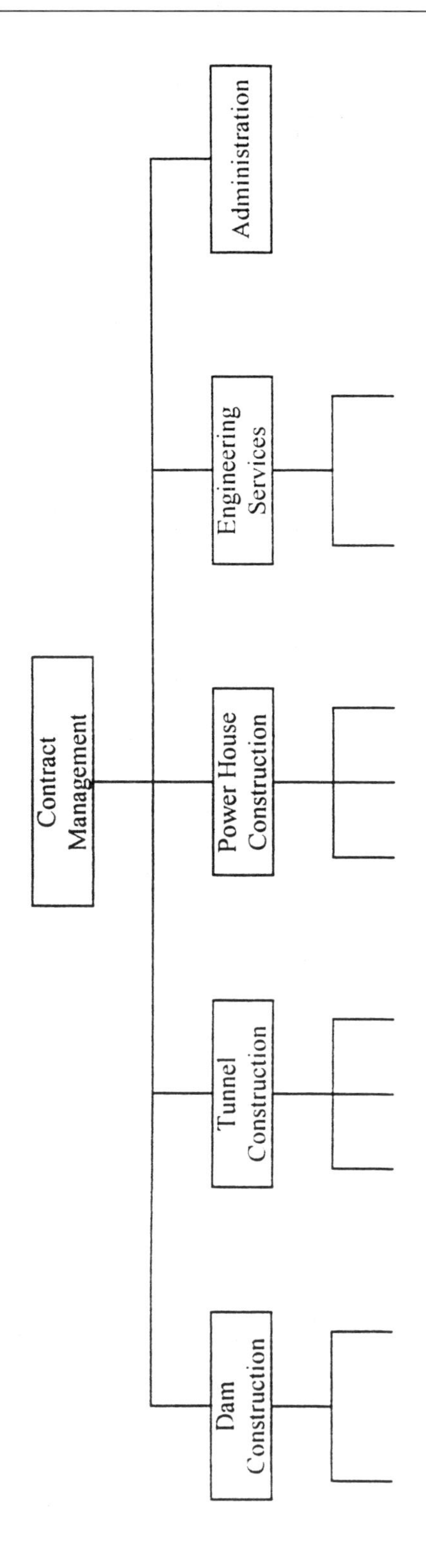

Figure 8.4 Construction site organization.

different specialists don't mix and so lack common interests and understanding. The risk is that none of them know how well their relationships are operating. Consequent failures in understanding and communications can be crippling when specialization is the basis of locating departments in different parts of a building, or more so if further apart. The loss of cooperation and of flexibility in the use of resources is likely to be uneconomic and demotivating. The consequences of one group or person making decisions independently of others' knowledge could also be unsafe.

The risks can be minimized by:

- minimum division of work;
- organization on the basis of the best work breakdown structure for each project, not vice versa;
- sustained attention to how the planned systems for communication and coordination are working.

8.5 Project teamwork

Success rarely occurs accidentally. Successful organization of work needs attention not only to what has to be done and who is able to do it, but also to how people are to relate together to produce what their projects need. This should include everyone who takes part in decisions on the objectives, scope, risks, standards, design, economy, financing, timing, methods, safety, control and acceptability of the projects. Anyone who represents a functional department's contribution to a project team should have the authority to commit their parent organization, so that decisions can be made collaboratively at the most knowledgeable level.

To concentrate on attention to project objectives and achieve good communications, everyone on a project should be located together. So why not instead organize work by project, i.e. employ all the people and other resources required for a project in a department directly under a leader responsible only for meeting the objectives of that project: forming what is known as a 'project task force'? That in turn should achieve communications, cooperation and motivation for each stage of a project, but in this also lie the disadvantages. One disadvantage is that it requires people to work themselves out of a job by completing their stage of the work. Another is that resources are not shared between projects, which can be wasteful at the time and fail to accumulate experience for the future.

Forming a dedicated team or 'task force' can bring together all that is needed to achieve the immediate objectives of a project, but at a longer-term cost that has traditionally been thought justified only for large, very urgent, one-off or unique projects. But these are relative words. As projects have become more risky, so has the need for flexibility combined with control. The dedicated team is agreed to be best for this.

A dedicated team located together may be demanded by the investors in a project because of their concern about a supplier's arrangements to complete a commitment. To the supplier it may appear to be an excessive immediate cost; it may result in the flexibility needed for survival. Reorganization is no solution to problems that are not due to organization. If a change is needed, then the formation of a separate project team involving all affected can be a means of resolving problems. Reorganization can utilize people's interest in the technical novelties or other features of a project to draw them into a changed system.

Further reading

Burns, T. and Stalker, G.M. (1986) *The Management of Innovation*, 2nd edn, Tavistock, London.

Conrath, D.W. (1968) The role of the informal organization in decision making in research and development. IEEE Transactions on Engineering Management, **EM-15** (3), 109–199.

Handy, C.B. (1985) *Understanding Organizations*, 3rd edn, Penguin, Harmondsworth.

Kakabadse, A. (1988) *Working in Organizations*, Penguin, Harmondsworth.

Kliem, R.L. and Ludin, I.S. (1992) *The People Side of Project Management*, Gower, Aldershot.

Martin, A.S. and Grover, F. (eds) (1988) *Managing People*, Thomas Telford, London.

Morris, P.W.G. and Hough, G.H. (1987) *The Anatomy of Major Projects*, Wiley, Chichester.

Turner, J.R. (1993) *The Handbook of Project-Based Management*, McGraw-Hill, London.

Wearne, S.H. (1993) *Principles of Engineering Organization*, 2nd edn, Thomas Telford, London.

Chapter 9

Project Management and Project Managers

The effectiveness of project management depends to a great extent upon the behavioural and managerial role of the key personnel, and in particular on the project manager. In large projects complex organizational systems may be needed for managing resources, but in a simple project one person can undertake all the internal managerial functions. The management of projects and the system of decision making is considered at the start of this chapter, leading into a review of matrix systems and matrix management.

9.1 Managerial hierarchies

Organizations are systems for managing resources. In a small organization one person can undertake all the internal managerial functions of planning, budgeting, organizing, motivating, monitoring and controlling the use of resources, and also the external managerial functions of buying, selling and reporting. Most organizations have grown to a size in which one person cannot do all this. We therefore usually find a hierarchy of managers in organizations, with delegation down from the 'top' through several levels of intermediate leaders. Every manager in the system is formally the link between the people immediately below him or her. Figure 9.1 shows an example.

A managerial hierarchy is a system of authority, essentially over expenditure and other decisions on resources. Decisions to take the risk of investing in a new project are usually made at a level higher than the subsequent responsibility for managing the consequences.

Managerial hierarchy is the normal basis of the organization of companies, public services, partnerships and cooperatives in all industrial countries, though with the difference that the manager at the top may be responsible to a board of directors, partners, parliament or all employees, with consequent differences in his or her autonomy.

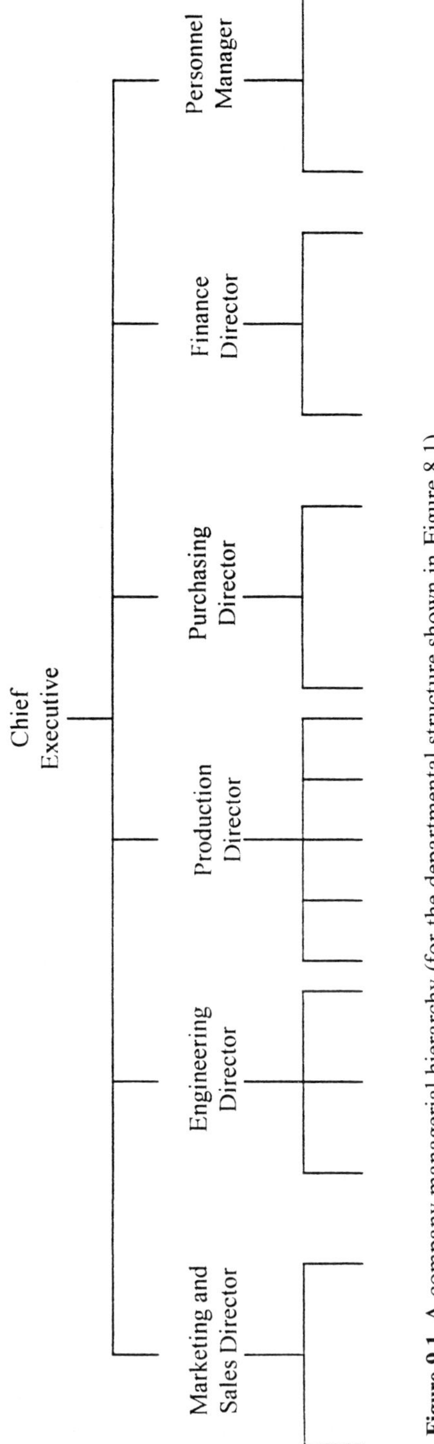

Figure 9.1 A company managerial hierarchy (for the departmental structure shown in Figure 8.1).

Figure 9.1 shows the hierarchy and job titles for the managers of the departmental structure previously shown in Figure 8.1, in Chapter 8. Figure 9.2 shows the managers for the example of a contractor's site structure previously shown in Figure 8.4. Organization charts or 'organigrams' like these are commonly used. The examples shown here are an illustration of the form of hierarchies. Job titles, the number of levels of management and the number of people responsible to each manager vary greatly with the size of organization, culture, and the nature of the work, but much of our language for discussing any system of organization is based upon this concept of authority stemming from the top of a pyramid-shaped hierarchy.

9.2 Levels of decisions

In classical organization theory the strategic or longer-term decisions are made at a high level in a managerial hierarchy, and these form the basis of more short-term local decisions at successive lower levels. Therefore, as proposed by Stafford Beer, a hierarchy should be a system of systems, each level having to think differently from the others to form a hierarchy of roles:

- directing
- exploring
- planning
- coordinating
- supervising
- producing.

Figure 9.1 indicates that each person has only one superior. In theory the sequence of decisions needed for a project could therefore be the basis for defining which manager should make each decision. Case studies show that systems in practice are more complicated than this, with multiple links between various levels and only indirect or occasional links between some managers and their immediate subordinates.

9.3 Management of projects

If everyone in an organization is employed on one project it seems obvious that the system can be simple, as the chief executive is also the project manager. This is rare, but can occur at the start of a new business, at a complete change in product, or at an amalgamation with another

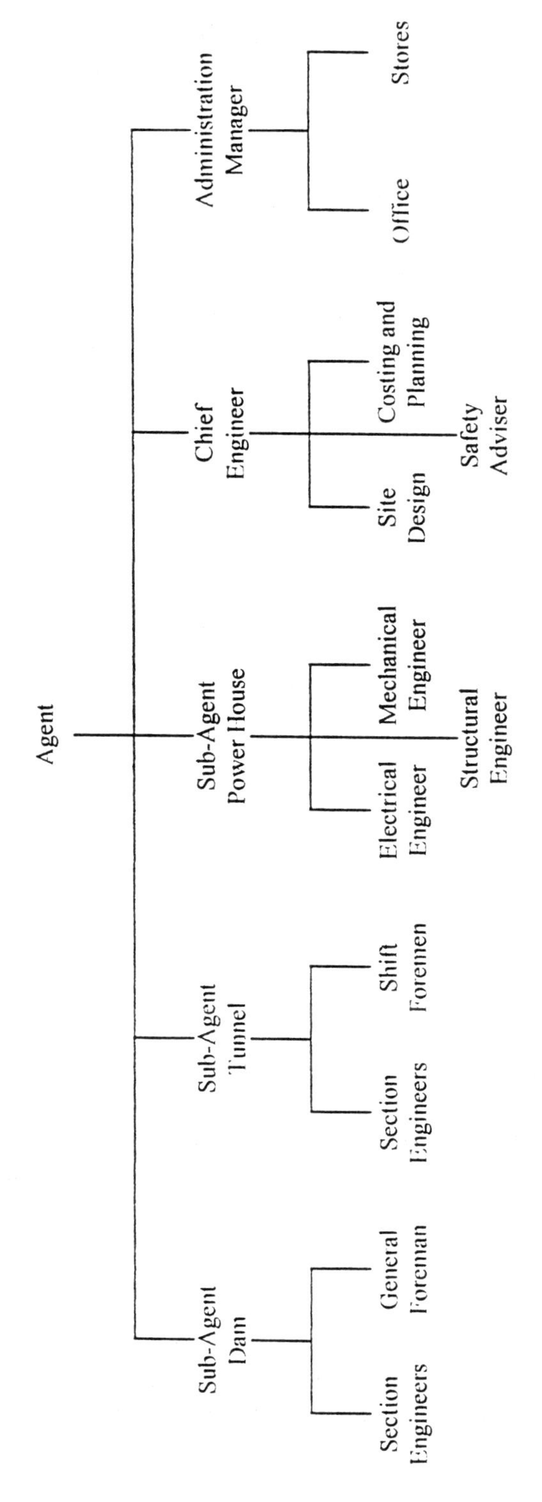

Figure 9.2 A site managerial hierarchy (for the sectional structure shown in Figure 8.4).

organization. A project to change all production in the organization shown in Figure 9.1 might be managed by the production director. Smaller or less vital projects can similarly be the responsibility of others at lower levels.

The traditional practice in manufacturing and construction firms has been that projects are the responsibility of departments in turn. In the organization shown in Figure 9.1, for instance, a technical proposal would initially be managed by the engineering department, a marketing proposal by the marketing department, and the initiators be responsible for consulting all others. The main risks of carrying out the resulting project would be mainly managed by the production department. This system works well if successive projects are similar, if priorities for the use of resources are agreed, and if the lessons of the results of the projects are applied by all the departments. A project planning and steering committee is a common way of achieving these conditions.

Practice has become more complex in all industries and public services to cope with the increase in the risks and changes needed during projects caused by greater uncertainties in predicting markets, technological innovations, competition and public controls. It is increasingly difficult for a project to be defined once and for all at the start and the detailed decisions delegated down a hierarchy of managers.

Project-dedicated teams

Some organizations have located together all the people required for a project, or at least those fully employed on a project during its design and subsequent stages. One instance is the placing of the design team for a construction project on site from the start of their work, incurring extra costs and requiring movement of personnel, but achieving concentration on the needs and risks of the project under one person, who is the project manager. Other instances include the creation of a complete design, development and manufacturing team away from an existing factory so that they concentrate on a new project.

Project coordination

For many projects, locating together all who need to know each other's thoughts and decisions may not be practical in order to share the resources with other projects. Or it may be unacceptable socially. If it cannot be done, a system of planning and coordination is needed to link everyone on a project.

One way of achieving this is to make one manager in the established hierarchy responsible for coordinating all decisions for a project, even though not the superior of all the people and other resources on that project.

Another is to appoint a project coordinator as an assistant to a manager who controls the resources required for the project. Figure 9.3 shows an example. In this the production director is the established 'line' manager in the hierarchy. The project coordinator is in what is called a 'staff' role. The staff role is in theory an extension of the line manager's capacity, not an additional level of management. One or a team of people can be in this position.

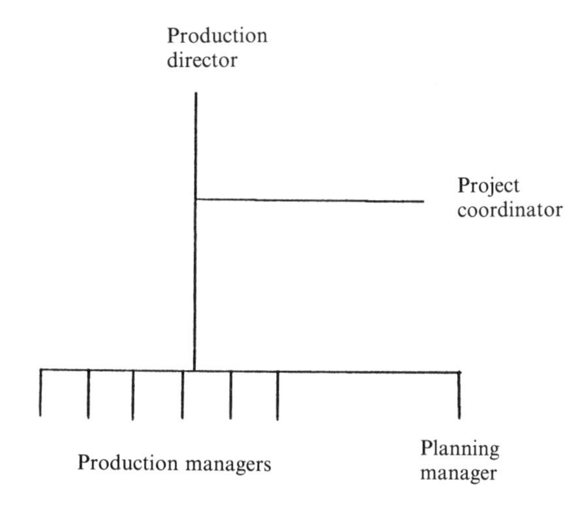

Figure 9.3 Project coordinator added to a managerial hierarchy.

The 'line-and-staff' arrangement works when everyone involved understands it and is willing to be influenced by a coordinator who is not their superior. This can be very dependent on the coordinator's personal qualities, expertise and skills. In times of conflict or uncertainty the role may be limited in its effectiveness because decisions affecting resources and commitments to others remain dependent upon the line members of the hierarchy of authority.

9.4 Matrix systems

To achieve leadership of each project and of each specialization used by the projects, firms and public authorities have evolved what are called 'matrix' systems of management, with separate roles for functional managers and project managers. Figure 9.4 shows an example where the resources of three departments are shared amongst three projects.

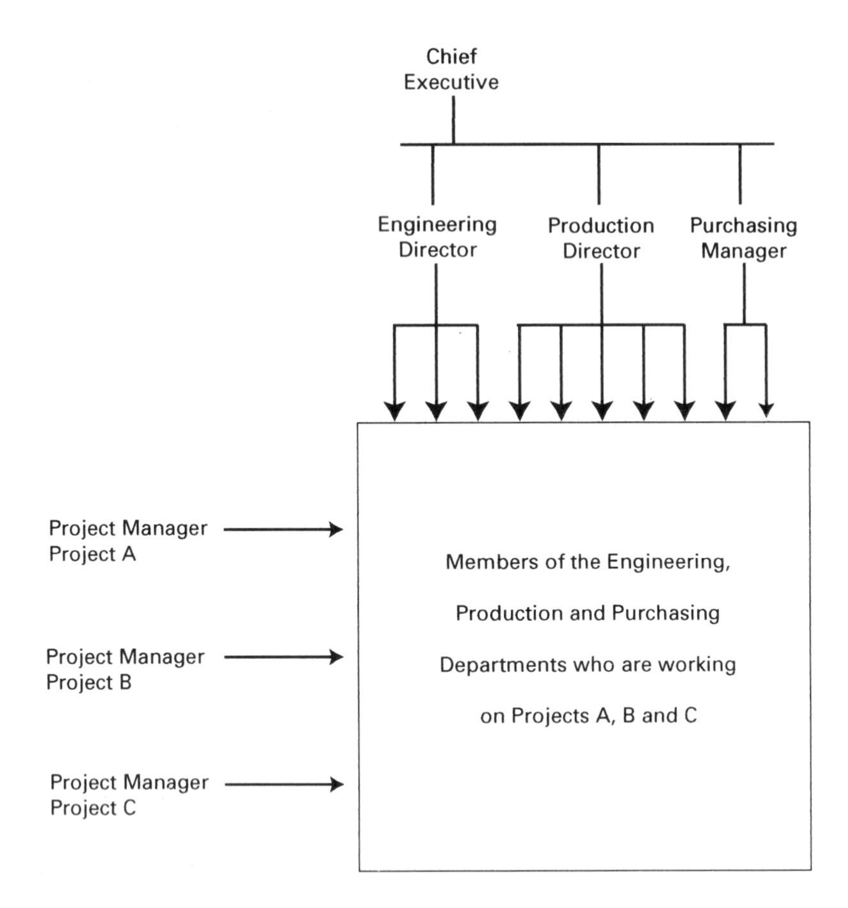

Figure 9.4 Matrix management of departments' resources.

Matrix systems provide opportunities to employ leaders with different skills and knowledge in these two types of managerial role, but the project and specialist managers should theoretically all influence decisions. The resulting systems can be complicated, particularly in the extent of the formal authority of the project managers relative to that of the functional managers.

The convention in organization charts is that the line of authority is vertical. Collaboration is thus horizontal. In a matrix these are compounded. Thus a diagram to show this needs to be drawn on the skew, as shown diagrammatically in Figure 9.5.

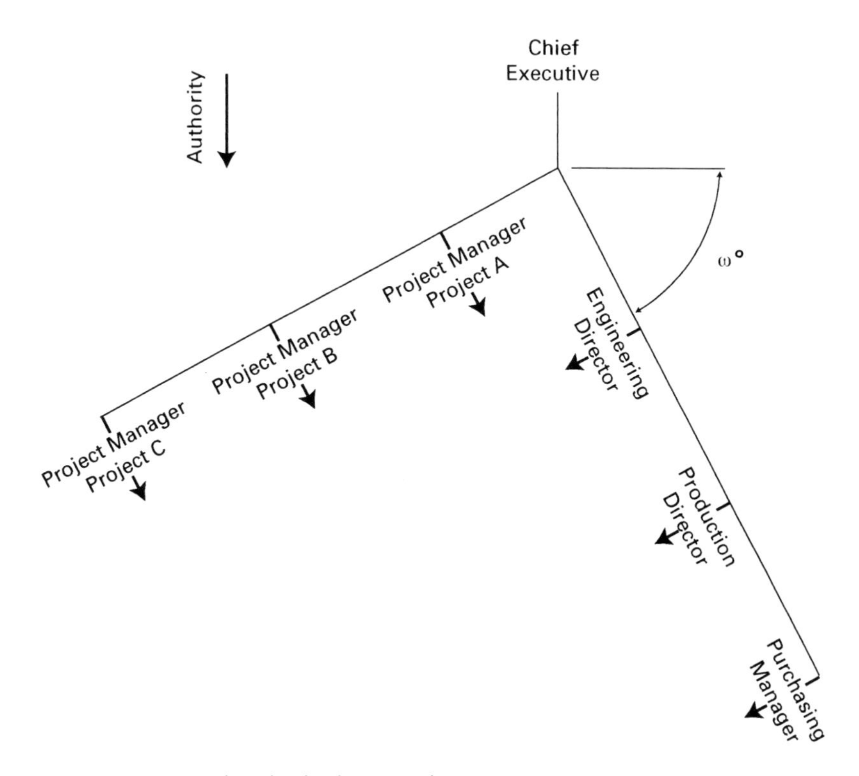

Figure 9.5 Formal authority in a matrix system.

The angle ω in effect indicates the relative influence of project and specialist responsibilities for achieving results. If ω is large, the system is called a *strong matrix*, meaning that the project managers have the greater influence over the quality and quantity of resources for their projects. In these cases, the specialist managers have to act as consultants to members of their sections once allocated to a project. A *weak matrix* is therefore one where the project managers are coordinators with little influence on resources.

In practice the angle ω is different from one project to another, depending upon the relative importance, urgency, uniqueness and size of each project. It also changes during the work for a project, depending upon the relative importance of innovation and control stage by stage.

In some examples a project manager is said to be responsible for the quantitative decisions affecting the cost and programme of his or her project, and the specialist managers responsible for qualitative standards in allocating people and other resources to each project. Formally this concept divides responsibilities for quality, cost and time. In the example shown in Figure 9.4 this potential problem can be avoided by defining that the project managers are responsible to the chief executive or to a steering committee, which represents all the departmental heads.

Three-dimensional matrix

Matrix systems in large organizations can be three-dimensional, for instance with these three arms:

- groups of specialists who design classes of subsystems for projects, for instance one designing transmission systems and another engines for a series of vehicles;
- groups of specialists who provide a service to the above groups, for instance metallurgists and vibration experts;
- project managers.

Managing in matrix systems

Matrix systems can work well given defined objectives and priorities for projects and with agreed amounts and quality of resources. They don't necessarily avoid conflicts over these. Examples indicate that their success depends upon:

- the will of the managers who control resources that their departments are to operate as services to projects;
- the personal skills and knowledge of the project managers;
- joint planning and decisions on priorities.

Relative responsibility

From case studies and published examples we can define at least nine degrees of relative project responsibility:

Chaser: the minimum project role – a person sent to report on problems after they have become apparent.

Monitor: an observer of progress and problems, reporting these to specialist or project managers.

Coordinator: as above, but expected to anticipate problems without authority over any work.

Planner: provider of planning advice and services, but with no control over project decisions.

Administrator: a coordinator with control over links between departments or between organizations.

Engineer: initiator of the project, possibly the systems designer, but otherwise a coordinator as above.

Controller: administrator with control over internal or contract payments.

Manager: combining all or most of the above, with a small team linking all work.

Director: the maximum project role – a manager executively in control of every person employed on a project.

The last of these categories consists of complete grouping into a project 'task force', the arrangement in an organization that is likely only for unique and urgent projects. Much more typical, but unfortunately much more complicated, is the tendency in organizations to learn by trial and error and start by appointing project chasers and find only later that they need to appoint project managers with the authority to try to avoid delays and other problems.

Coordinating roles can be most difficult to make effective when first added to a system. In studying such roles Argyris observed that the position is one of power based upon what ought to be the best knowledge of the needs of a project. If there is no policy for the project, the project coordinator must create one. If there is a lack of decisions, he/she must make them. The project coordinator is the 'champion' for the project.

9.5 Matrix management or internal contracts

Observations of matrix systems indicate that there can be conflicts in them between project managers and the heads of specialist groups about the allocation of resources to a project and the quality, cost and timing of the work to be done by them. The heads of the functions should earlier have agreed on specifications, budgets and programmes for every project, but may have done so some time before a project starts and then only in sufficient detail to get a budget or a contract to proceed. This may not ensure that adequate resources are available when a project calls for them.

One means of avoiding most or all such problems is to treat each specialist group's work for a project as an internal 'contractual' commitment. If their work was to be purchased by contract from another organization there would normally be a prior process of investigating the potential supplier's capacity and understanding of the work required, followed by an invitation to offer to do it for a price and specified quality and delivery. For critical or risky work, procedures would also be agreed for progress reporting, inspection, changes and resolving problems before they might be needed. The same are in effect needed within organizations, not through legally enforceable documents but by agreeing definitions of what is expected of others rather than assuming that these are understood and agreed.

External commitments

The conduct of contractual links between a customer and a supplier is traditionally delegated at the making of a contract, in both organizations. Up to that point the links with the other party are the responsibility of the level of manager authorized in an organization to enter into the contract. Traditionally once the contract has been made the authority to communicate with the other party is delegated, typically two levels down as shown in Figure 9.6. This is to be seen in parties to contracts in all industries. Entering into a contract is expected to be much riskier than carrying the obligations in the contract. Delegation as seen in these cases follows the principle of levels of management being authorized to take levels of risk.

Delegation two levels down is logical in both the buying and the selling parties' organizations if they are in their usual situation of having several contracts for one or more projects. A manager authorized to enter into a contract delegates the continuing commitments to a manager below him. In turn that manager delegates that particular contract to a subordinate, unless it is exceptionally risky or important.

The traditional principle common to the buying and selling organizations is that of delegation of authority appropriate to the risks to their organization in the decisions to be made. The pattern is that commitments are entered into by decisions involving relatively big risks, and that the remaining decisions are matters of detail and lesser risks. The basis of this is that risks reduce as a project proceeds, in a big step at the point of entering into the contract. Thus the logical hierarchy of decisions has corresponded to the classical hierarchy of management.

Contract administration

The result of the above is that both organizations are represented in relationships with the other through an individual responsible for the links between them which it is expected will be needed in carrying out both parties' obligations in the contract.

The common description is that these individuals are responsible for *contract administration*. In this role each is in charge of something they

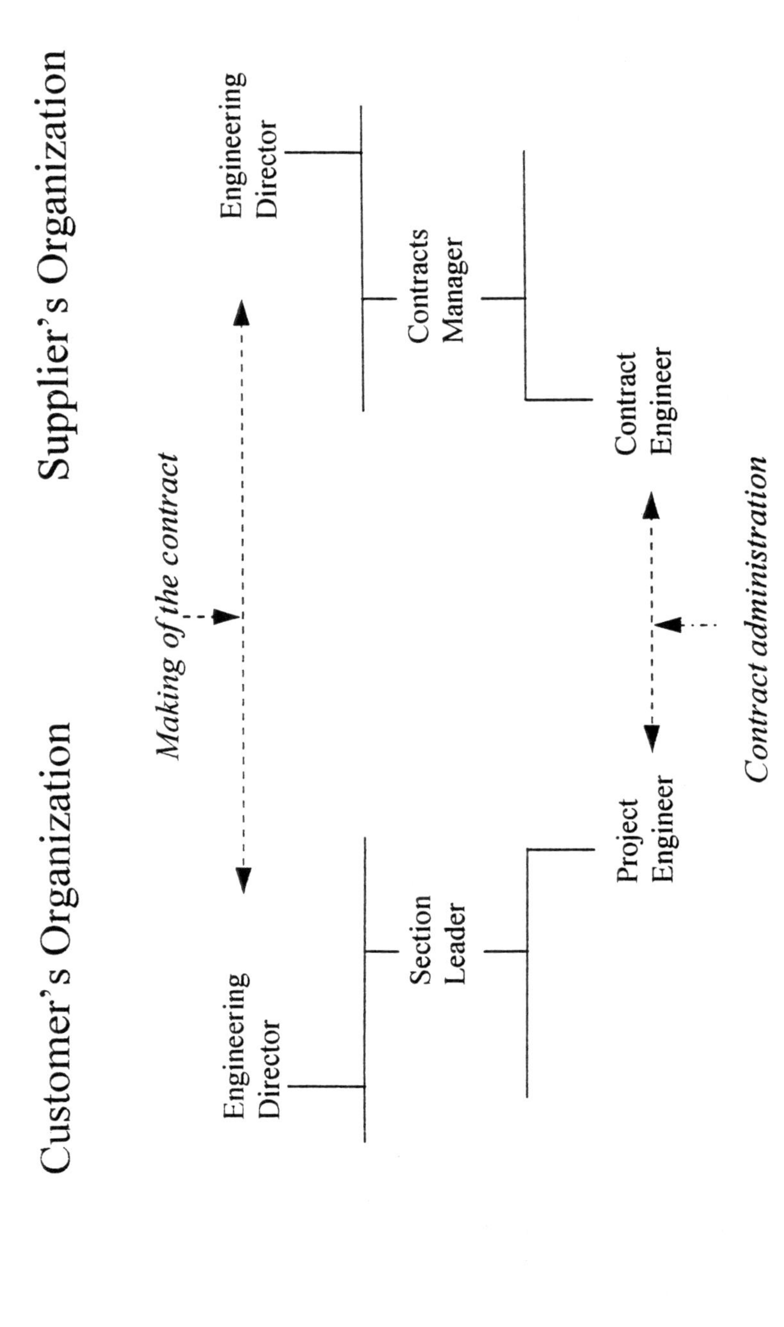

Figure 9.6 Hierarchical delegation of contractual relationships.

do not own, do not control, did not create, and in practice may not have been much involved in agreeing. The terms of the contract may include a procedure that they operate for changes to the scope, timing and costs of the work, but they are not authorized to renegotiate or terminate the contract. That authority remains at the level of those authorized to enter into that contract.

Thus using the job titles shown in Figure 9.6, the supplier's contract engineer has little or no formal authority within his organization to apply resources to the project, and the customer's project engineer usually has only a limited 'contingency' amount in his budget for extra payment to the supplier for changes. Achieving successful results from delegating contract administration therefore depends upon the higher managers' foreseeing the risks and budgeting for them at the time of entering into a contract, plus the ability of the project engineer and the contract engineer to anticipate the unexpected and to influence the managers in their organizations who have the formal authority over budgets and resources.

In practice the relationships within an organization are usually more complex than described above. There can be much consultation between various levels of managers and advisers before decisions, so that the role of the manager described as having the authority to make a contract may be more that of leader of discussions and decisions. What is commonly agreed as desirable is that one person should be the channel of communications with the other party to a contract.

Managing contract risks

A customer's risks do not reduce appreciably when work is ordered from a supplier, though under most contracts the suppliers are committed to achieving performance in terms of quality, schedule and support services. The success of such contracts depend on a customer's decisions in specifying the goods and services to be supplied being sufficiently accurate and final; any subsequent changes can be incorporated without reversing progress in completing the contract commitments. If they are not complete and final, communications with each supplier have to continue to be the responsibility of a manager who has the authority to agree the cost and other effects of the risks of changes.

The success of a contract also depends on the chosen supplier obtaining good performance from his subcontractors, yet however well selected and subject to the customer's checking the cost risks and price competition for subcontracts make the performance and survival of subcontractors less and less predictable. The wise customer's project manager is therefore given the time and the authority to see that a supplier is obtaining planning and other information from subcontractors to ascertain that they will deliver right first time on time.

Further reading

Argyris, C. (1967) Today's problems with tomorrow's organizations. *Journal of Management Studies*, **1**, 31–35.

Beer, S. (1972) *The Brain of the Firm*, Allen Lane and Penguin.

Burbridge, R.N. (ed.) (1988) *Perspectives on Project Management*, Peter Peregrinus (Institution of Electrical Engineers), London.

Cleland, D.I. and Gareis, R. (1993) *Global Project Management*, McGraw-Hill, New York.

Gaisford, R.W. (1986) Project management in the North Sea. *International Journal of Project Management*, **4** (1), 5–12.

Leech, D.J. and Turner, B.T. (1990) *Project Management for Profit*, Ellis Horwood, Chichester.

Lock, D. (ed.) (1992) *Project Management Handbook*, 2nd edn, Gower, Aldershot.

Martin, A.S. and Grover, F. (eds) (1988) *Managing People*, Thomas Telford, London.

Merna, A. and Smith, N.J. (1990) Project managers and the use of turnkey contracts. *International Journal of Project Management*, **8** (3), 183–189.

Morris, P.W.G. (1989) *Managing project interfaces*. Technical Paper no 7, Major Projects Association, Oxford.

Nochur, K.S. and Allen, T.J. (1992) Do nominated boundary spanners become effective technological gatekeepers?, *IEEE Transactions on Engineering Management*, **39** (3), 265–269.

Rutter, P.A. and Martin, A.S. (eds) (1990) *Management of Design Offices*, Thomas Telford, London.

Thompson, P.A. (1991) The client role in project management. *International Journal of Project Management*, **9** (2), 90–92.

Wearne, S.H. (1994) Preparing for privatized project management. *International Journal of Project Management*, **12** (2), 118–120.

Part II
Project Operations

Chapter 10
Planning

Planning has been accepted as an important function in the process of project management. Without concentrating on mathematical analysis in detail, this chapter reviews programming and planning techniques for design, procurement, construction and commissioning of projects. The role of information technology (IT) is examined together with suggestions on the identification and selection of appropriate software packages.

10.1 Planning

The successful realization of a project will depend upon careful and continuous planning. The activities of designers, manufacturers, suppliers, contractors and all their resources must be organized and integrated to meet the objectives set by the promoter and/or the contractor.

The purposes of planning are to persuade people to perform tasks before they delay the operations of other groups of people, and in such a sequence that the best use is made of available resources; and to provide a framework for decision making in the event of change. Assumptions are invariably made as a plan is developed: these should be clearly stated so that everyone using the plan is aware of limitations on its validity. Programmes are essentially two-dimensional graphs and in many cases are used as the initial, and sometimes the only, planning technique.

Packages of work, usually referred to as *activities* or *tasks*, are determined by consideration of the type of work, or the location of the work, or by any restraints on the continuity of the activity. Activities consume *resources*, which are the productive aspects of the project and usually include the organization and utilization of people (labour), equipment (plant) and raw material. Sequences of activities will be linked on a time-scale to ensure that priorities are identified and that efficient use is made of expensive and/or scarce resources, within the physical constraints affecting the job.

A degree of change and uncertainty is inherent in engineering, and it should be expected that a plan will change. It must therefore be capable of being updated quickly and regularly if it is to remain a guide to the most efficient way of completing the project. The plan should therefore be simple, so that updating is straightforward and does not demand the feedback of large amounts of data, and flexible, so that all alternative courses of action can be considered. This may be achieved either by allocating additional resources or by introducing a greater element of float and extending the contract duration, as necessary. In either case, the estimated cost will increase: hence it is essential to link the programme with the cost forecast.

It is difficult to enforce a plan that is conceived in isolation, and it is therefore essential to involve the people responsible for the constituent operations in the development of the plan. The plan must not impose excessive restraint on the other members of the organization: it should provide a flexible framework within which they can exercise their own initiative. The plan must precipitate action and must therefore be available in advance of the task.

10.2 Programming

Programmes are required at various stages in the contract: when considering feasibility or sanction, at the pre-contract stage and during the contract. They may be used for initial budget control or for day-to-day construction work. They may pertain to one contract, or a number of contracts in one large project.

The planner must therefore decide on the appropriate level of detail for his programme and the choice of programming technique. Important factors in the choice include the purpose of the programme, the relevant level of management and the level of detail required. Simplicity and flexibility are the keys; a programme of 100 activities is easy to comprehend whereas a programme of 1000 activities is not. Often it is good practice to ensure that the number of individual activities should relate closely to the basic packages of work required, or to the cost centres defined in the estimate, and should all have durations of a similar order of magnitude.

The period of time necessary to execute the work of an activity, the *duration*, depends on the level of resources allocated to the activity, the output of those resources and the quantity of work. The duration may also depend on other outside restraints, such as the specified completion date for the whole or some part of the work, the delivery date for specific material or restrictions on access to parts of the works.

A number of common forms of programme used in engineering project management are reviewed below.

Bar charts

The most common form of plan is the bar chart, also known as the *Gantt chart*: an example is shown in Figure 10.1. Each activity is shown in its scheduled position to give efficient use of the resources; the logic and float are shown by dotted lines and boxes; and important constraints or key dates are clearly marked. The space within the bars can be used for figures of output or plant costs, and there is room beneath to mark actual progress. Frequently it is useful to plot a period-by-period histogram of the demand for key resources directly under the bars at the bottom edge of the programme.

Line of balance

This simple technique was developed for house building, and is useful for any repetitive type of work. The axes are the number of completed units and time; the work of each gang appears as an inclined line, the inclination being related to the output of the gang, as shown in Figure 10.2.

Location–time diagram

In cross-country jobs such as major roadworks, the erection of transmission lines, or pipelaying, the performance of individual activities will be greatly affected by their location and the various physical conditions encountered. Restricted access to the works, the relative positions of cuttings and embankments, sources of materials from quarries and temporary borrowpits, the need to provide temporary or permanent crossings for watercourses, roads and railways, and the nature of the ground will all influence the continuity of construction work and the output achieved by similar resources of men and machines working in different locations (Figure 10.3).

10.3 Network analysis

There are two basic forms of network analysis techniques: *precedence diagrams* (sometimes called activity-on-node networks) and *arrow diagrams* (sometimes called activity-on-line networks).

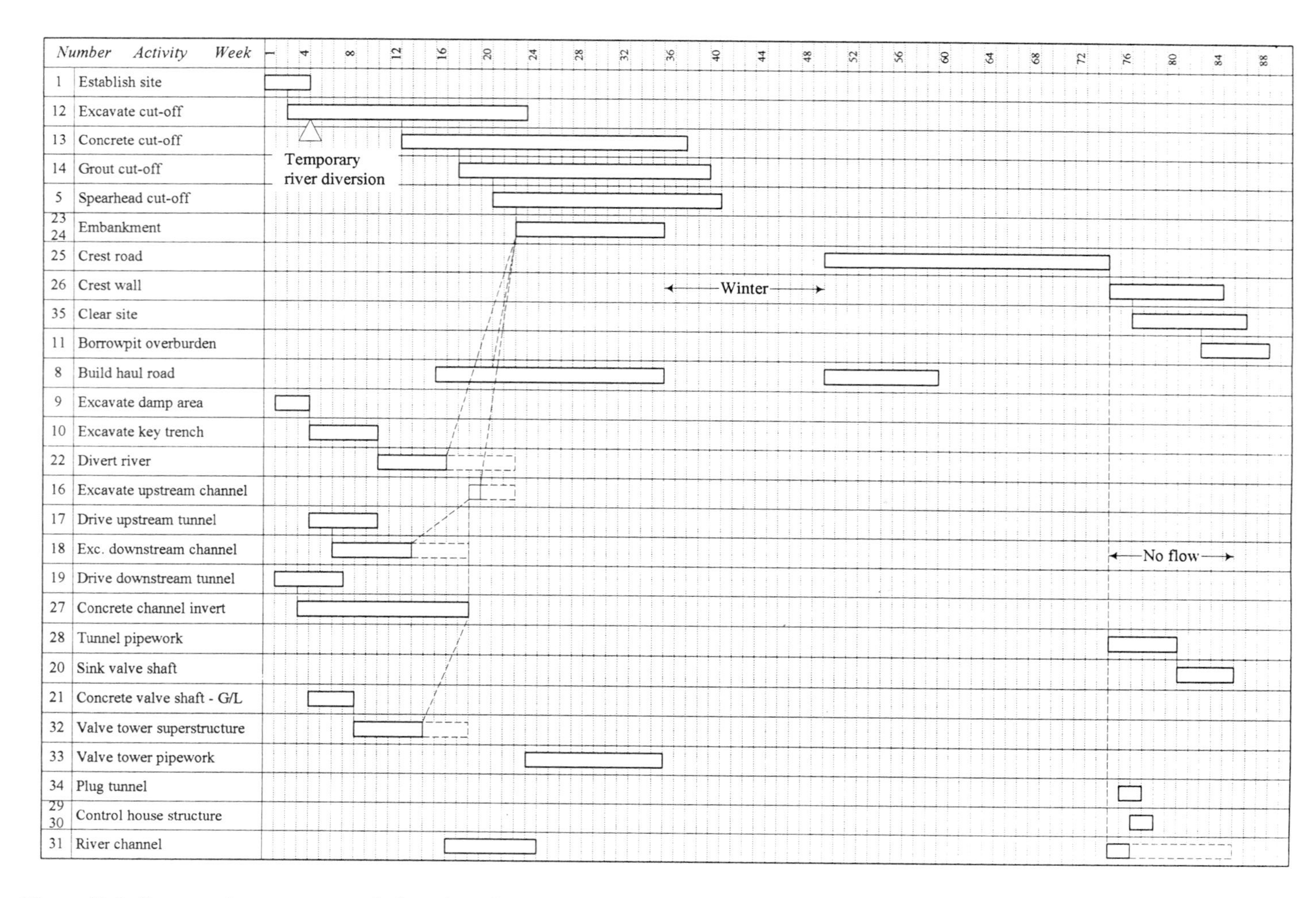

Figure 10.1 Construction programme in bar chart form.

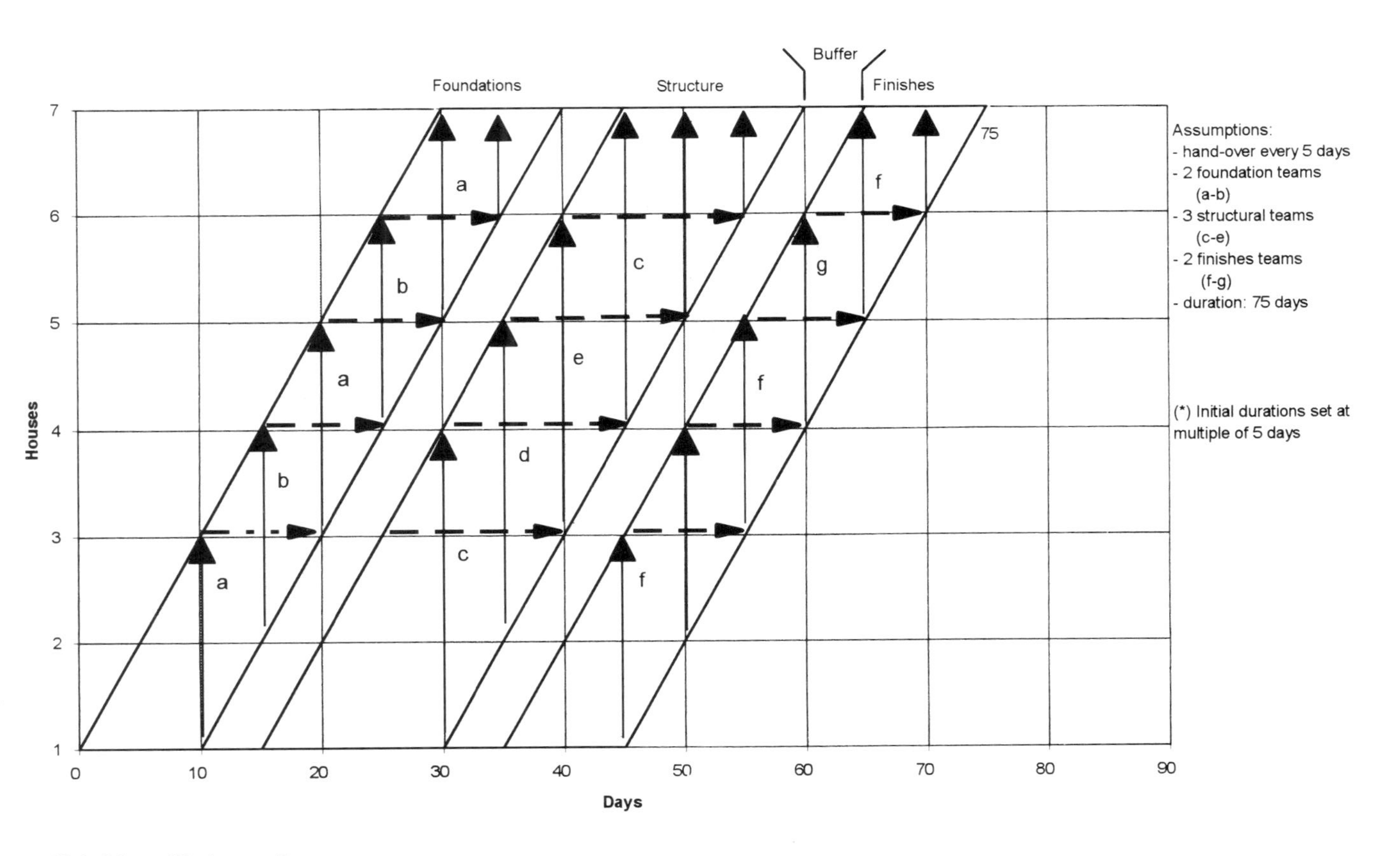

Figure 10.2 Line of balance diagram.

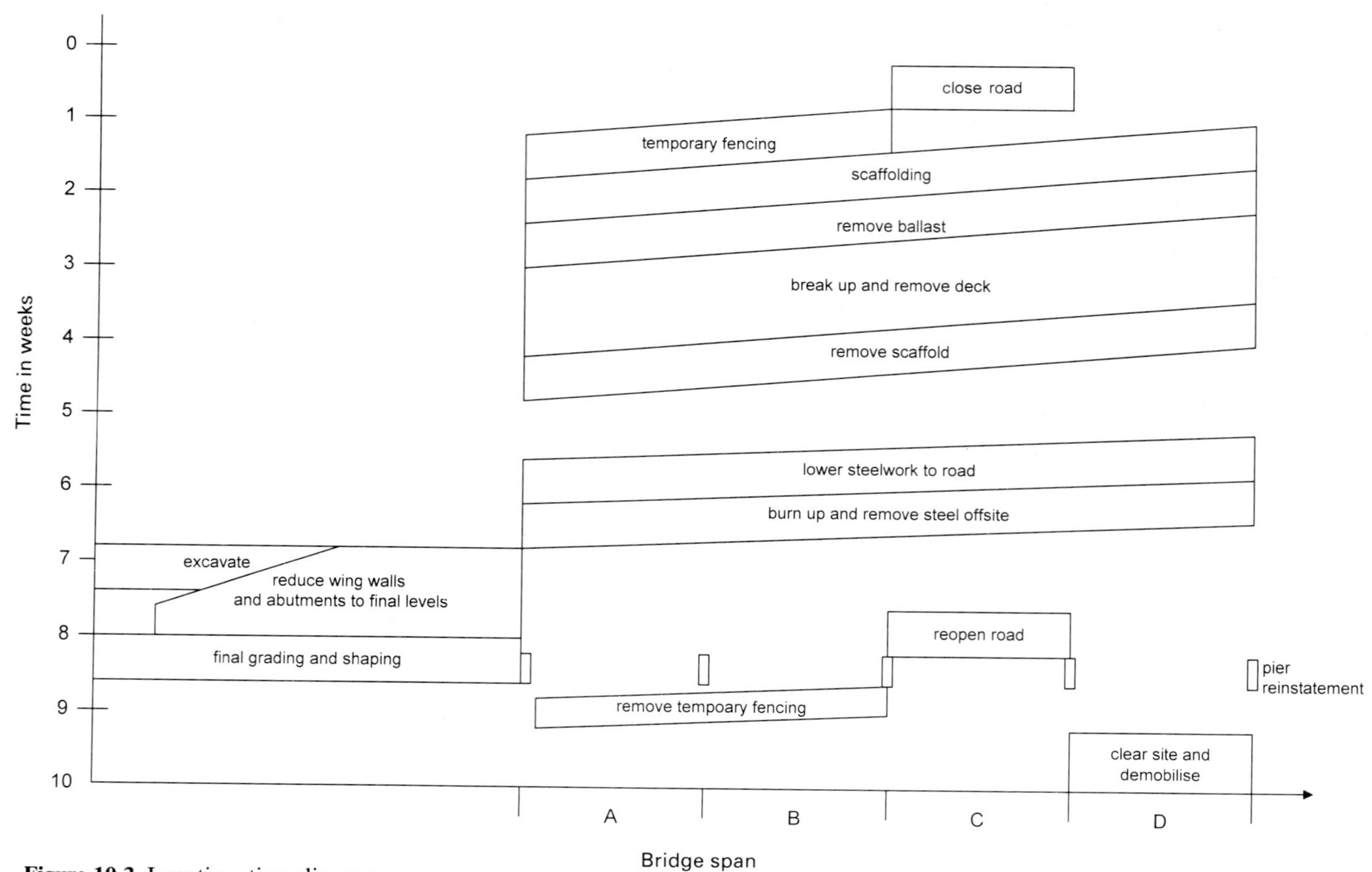

Figure 10.3 Location–time diagram.

Although both methods will achieve the same result, the precedence diagram is preference. The advantages of the precedence system over arrow diagrams may be listed as follows.

- They are flexible; logic is defined in two stages.
- Dummy activities are eliminated.
- Revision and introduction of new activities is simple.
- Overlapping and delaying of activities is easily defined.
- The use of preprinted node sheets is possible.

The network diagram (Figure 10.4) resembles a flow chart: activities are represented by box-shaped nodes and the interrelationships between activities by lines known as *dependences*. The dependences are developed by moving to each activity in turn, asking 'What can start once this activity has started?' and drawing the relevant lines. The convention is that dependencies run from nose to tail of succeeding activities: that is, from the finish of one activity to the start of the succeeding activity. Figure 10.4 relates to the following worked example.

Apart from the *critical path* a degree of choice exists in the timing of the other activities – a characteristic that is called *float*. It is the float that will be utilized later to adjust the timing of activities in order to obtain the best possible use of resources. The *total float* associated with an activity is the difference between its earliest and latest starts or finishes.

Free float is the minimum difference between the earliest finish time of

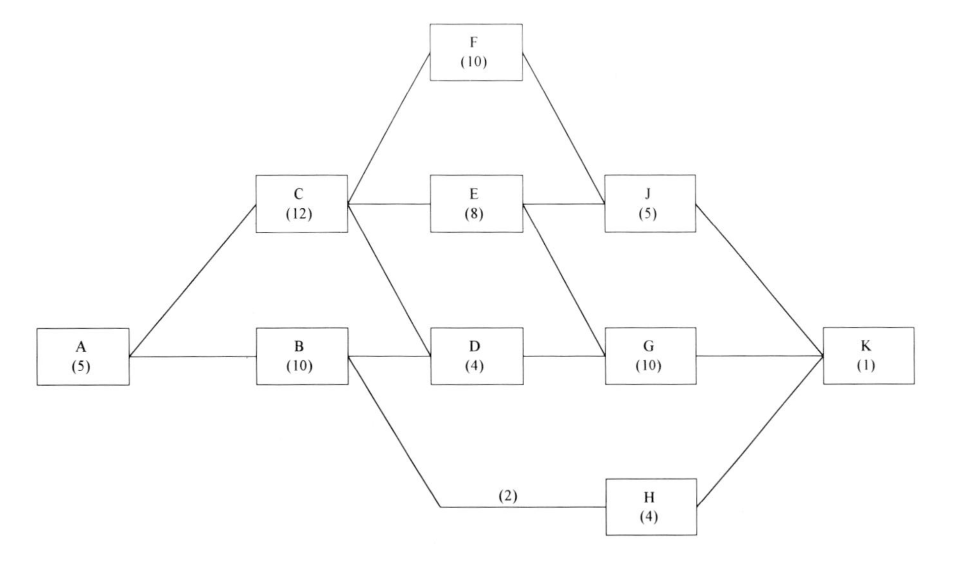

Figure 10.4 Precedence diagram for network analysis.

that activity and the earliest start time of a succeeding activity. Total float is a measure of the maximum adjustment that may be made to the timing of an activity without extending the overall duration of the project; free float is that part of the total float that can be used without affecting subsequent activities.

To illustrate the techniques needed to perform the precedence diagram critical path method a worked example based on a simple ten-activity network is described below.

Worked example

Table 10.1 shows the durations and interdependence of ten activities required to carry out a project.

(a) Construct a precedence diagram assuming no resource restrictions and calculate the minimum duration of the project.
(b) Schedule the earliest, latest start and finish for each activity and show the critical path.
(c) If activity F is extended to a duration of 15 days, what is the effect on the critical path?

Table 10.1 Network example.

Activity	Duration (days)	Logic
		Depends only on the completion of:
A	5	–
B	10	A
C	12	A
D	4	B and C
E	8	C
F	10	C
G	10	E and D
H	4	B [2 days overlap]
J	5	E and F
K	1	H and G and J

Solution

(a) Start with the first activity A, at the left-hand edge in the centre of the preprinted sheet. As a start activity it has only successor activities. Using the logic in Table 10.1, the start of activities B and C is dependent upon the completion of activity A, so these activities are plotted immediately adjacent to A. This process is repeated until activity K is plotted. The network should then be checked to ensure that there are no loops of activities and that there are no unconnected strings of activities (Figure 10.4).

The next step is to complete the forward pass. The project will commence at the end of time period zero. Hence the early start for activity A will be 0. The early finish is calculated by adding the activity duration to the early start: that is, $0 + 5 = 5$. The early finish of A is then the early start of both B and C, day 5, and the process is repeated. Activity D has two predecessor early finishes from activity B and activity C; as D cannot start until both B and C are completed the larger value for the early finish from C, day 17, becomes the early start for D.

The link between activity H and activity B is subject to an overlap of two days: that is, the start of H can overlap the finish of B by two days. The early finish for B is day 15 and so the early start for H is day 13. The forward pass continues until activity K is reached and an early finish for the project is calculated as 36 days.

(b) The process is then reversed to calculate the backward pass. Using the same logic but starting at activity K a late finish is determined. In this example, as in the majority of cases, the late finish is taken to be the same as the early finish, day 36. Note that it is possible to insert a higher number as the late finish but the backward pass would then not end at time period zero. By the same logic the late start is calculated by subtracting the activity duration from the late finish. This process is repeated, taking the lower value on the backward pass and adding an overlap until a late start for activity A is calculated. Unless this value is also zero an error has been made.

Each activity will now have an earliest and latest start and an earliest and latest finish. A number of activities, including the start activity and the finish activity, will have identical early starts and late starts. The difference between the late start and the early start is known as float. The critical path or paths consists of the activities with zero float logically linked between the start activity and the finish activity. This gives a critical path of A–C–E–G–K.

(c) Activity F is not critical; it has a float of 3 days. By extending the duration by 5 days, the early finish of activity F is extended by 5 days also. The forward pass is continued to show that activity K now finishes at day 38: an extension of 2 days. When the backward pass is completed it is shown that there is a new critical path and changed float durations for non critical activities. The new critical path is A–C–F–J–K.

10.4 Updating the network

Updating should be done frequently to ensure that the network is relevant. The procedure is as follows.

(1) Identify and mark all activities on which work is currently proceeding as 'live' activities. This is important, as it focuses attention on the future. What has happened to the present date must be accepted and concentration given to replanning and scheduling future activities.
(2) Introduce a new activity at the origin of the programme, having a duration equal to the time interval between the start of the programme and the date of updating.
(3) Change the durations of all completed activities to zero.
(4) Calculate revised durations from the date of update for all live activities and all future activities, taking into account any changes in requirements or actual performance. Note that a completely new estimate of the amount of work remaining to be done should be made for each 'live or future' activity at each update. The revised activity duration is derived from this figure and a reassessment value of the probable output of the relevant resources.
(5) Evaluate the programme in the normal way.

10.5 Resource scheduling

In the initial stages of planning a project a precedence network has been constructed. The data for this have usually been derived using the most efficient sizes of gangs and the normal number of machines, with the assumption that the resource demands for each activity can be met. Bar charts have then been drawn using earliest start dates and showing maximum float.

The next step is to consider the total demand for key resources. When considering the project or contract as a whole, there will be competition between activities for resources, and the demand may either exceed the planned availability or produce a fluctuating pattern of their use. This is known as *resource aggregation.*

The next stage, usually known as *resource levelling* or *smoothing*, utilizes the project float. Float can be used to adjust the timing of activities so that the imposed resource limits and the earliest completion date are not exceeded. If the float available within the programme is not sufficient to adjust the activities, the planner could consider the resources given to each activity and assess whether the usage can be changed, so altering the individual activity durations. It is clear that the levelling of one resource will have an effect on the usage of others. In consequence, resource levelling is usually only applied to a few key resources.

In some cases it will still not be possible to satisfy both the restraints on resource availability and the previously calculated earliest completion date, and the duration of the project is then extended. The lower the limits placed on resources, the greater the extension of the completion period of the project when it is 'resource scheduled'. Once the key resources have been adjusted, a new completion date results. If this is not acceptable, the resource limits must be adjusted and the process repeated. When resource scheduling has produced a satisfactory solution, the start and finish dates for each activity are said to be their *scheduled* values. It is probable that few scheduled activities will offer float.

10.6 Planning with uncertainty

The planner is often faced with a degree of uncertainty in the data used for planning. Most planning techniques inevitably use single-point, deterministic, values for duration and cost, though in practice there may be a range of values for these parameters.

There may be times when it is more appropriate to consider the uncertainties of the project. There are a number of ways in which this can be achieved, ranging from the simple techniques of using pessimistic, optimistic and most likely three-point values, through sensitivity analysis, to the more sophisticated Monte Carlo random sampling method. In all cases it causes the planner to adopt a more statistical approach to the data, and it focuses attention on the uncertainties in the project.

When using such techniques the planner is probably forced into using a computer. The mechanics and logistics of multivalue probabilistic models eliminate manual calculation on all but the simplest of construction programmes.

One planning technique that does not necessarily require such powerful computation tools is the *decision tree approach.* Used as a means to judge between decisions, this technique can be of assistance in the early stages of planning. Full details of the technique can be found in many standard texts.

10.7 Software and modelling

The increasing commercial pressure to achieve predetermined time and cost targets, combined with the power of the desktop computer, has led to a proliferation of project management software packages. These programs vary widely in terms of their modelling flexibility and simulation options, but are designed to serve the same purpose, which is to provide project managers with the power to plan the time and cost outturn in projects. All programs link time, cost and the resources of the project and allow the project manager to interactively forecast the financial commitment for the project and to assess a range of scenarios to reflect likely change and uncertainty.

Most project management software is based upon some form of mathematical model, most commonly a network. The project manager is concerned with three key factors: time, cost and quality. As quality can largely be controlled by specifications and contractual procedures, the mathematical model is required to represent the interrelationship between the other two factors. The model is based upon a real project, often the client's estimate, and then simulation, the changing of the model, is used to try to predict the future outcome of the real project.

Wherever possible it would seem desirable to be able to use data that have to be assembled as part of the overall requirements of the project to be used as input data for the software program used to manage the project. However, this is not always possible, and it is important to be aware of the significance of simplified or modified input data for a time and cost control package.

Project appraisal

This is a process of investigation, review and evaluation undertaken as the project or alternative concepts of the project are defined. A project model in this project stage is dominated by uncertainty and with little information likely to be available should be a simple network model with no more than 50 activities. The program should meet the following main needs:

- network modelling facilities
- the flexibility to spread costs and resources
- the discounted cash flow technique (DCF)
- a flexible risk-modelling module
- reporting facilities for the simulation results
- export/import options.

Construction/erection phase

This requires a realistic plan that sets out logically how the works are to be built, includes some contingency for unexpected events and sets achievable targets against which progress can be monitored. The construction stage in general requires a detailed model. The activity network, with costs and resources allocated, should be sensitive to the impact of changes made to time and cost components, so that action can be taken to prevent delays and additional costs. The program should meet the following criteria:

- detailed network-modelling facilities
- detailed cost- and resource-modelling facilities
- intelligent resource-scheduling options
- progress measurement, 'earned value principle'
- a comprehensive customizable report module
- export/import options.

10.8 Selecting project management software

Evaluating programs by using them to model and to simulate projects is in most cases very time-consuming and inefficient. Typically the project manager could be asked to base a selection on a range of brochures and demonstration versions of project management packages. Mistakes can be made at this stage, and a rational selection procedure is needed.

The selection of project management software needs to be based on both subjective and objective information. A method of integrating this information is to use *decision analysis*, which permits quantitative evaluation of the various courses of action, to find the best possible program based on the project manager's needs. Decision analysis provides not only the philosophical foundations, but also a logical and quantitative procedure for decision making.

The program selection system outlined here is based on a decision table, where the project manager lists the needs, then tests software packages and gives the programs points. The points are added and the most feasible program should then be the one with the highest program rate, as illustrated in Figure 10.5.

The proposed system uses superfactor 1 as 'appropriateness of program' and superfactor 2 as 'cost'. 'Appropriateness of program' consists of the following main subfactors: user-friendliness, modelling flexibility, simulation options, reporting facilities and export/import options. 'Cost' includes the purchase price, the costs of learning the program, and the maintenance costs. The appropriateness of program was given superweight 5 and cost superweight 2. The subfactors are

	Superfactors	Superfactor 1	Superfactor 2
	Superweights	1–5	1–5
	Subfactors	Subfactor 1	Subfactor 2
	Subweights	1–5	1–5
	Subfactor rate	1–5	
	A	Subweights × subfactor rates	
	B	sum A/sum subweights	
Program A	Program rate	Sum (B × superweights)	

Figure 10.5 Network example.

found by breaking the superfactors into modules, and the subfactors are given subweights between 1 and 5. A program rate can then be calculated. Figure 10.6 shows results from applying the system on the three programs based on appraisal stage modelling and simulation criteria.

Large companies with considerable in-house computing experience very often have specialist integrated programs for performing project management tasks, which would prove too expensive for smaller organizations. One of the future challenges for software producers is to create flexible programs for planning, estimating, control, accounting, procurement and risk analysis with common interfaces. The trend though seems to be to impress users by giving emphasis to the user interface and the screen graphics by creating endless view-screens and numerous ways of inputting data, instead of increasing the time- and cost-modelling flexibility, simplifying the simulation options, and improving program communication and program heuristics.

Further reading

Antill, J.M. and Woodhead, R.W. (1982) *Critical Path Methods in Construction*, 3rd edn, John Wiley & Sons, Chichester.

Chapman, C.B., Cooper, D.F. and Page, M.J. (1987) *Management for Engineers*, John Wiley & Sons, Chichester.

Smith, A.A., Hinton, E. and Lewis, R.W. (1983) *Civil Engineering Systems*, John Wiley & Sons, Chichester.

Thompson, P.A. (1981) *Organization and Economics of Construction*, McGraw-Hill, London.

	Superfactor	Appropriateness of program									Cost		
	Superweight	5									2		
	Subfactor	Start up	Use	Network	Resource	Cost	Updating	Reporting	Risk	Export/ import	Buy	Learn	Maintain
	Subweight	2	4	5	4	4	4	5	0	3	5	3	3
Program I	Subrate	4	4	5	4	3	5	4	0	3	3	2	2
	A	8	16	25	16	12	20	20	0	9	15	6	6
	B	126/31 = 4.1									27/11 = 2.5		
	Program rate	25.5											
Program II	Subrate	4	4	3	3	2	3	3	0	2	4	3	2
	A	8	16	15	12	8	12	15	0	6	20	9	6
	B	92/31 = 3.0									35/11 = 3.2		
	Program rate	21.4											
Program III	Subrate	3	3	2	2	5	0	2	0	1	3	2	3
	A	6	12	10	8	20	0	10	0	3	15	6	9
	B	69/31 = 2.2									30/11 = 2.7		
	Program rate	16.4											

Figure 10.6 Selecting project management software.

Exercises

Suggestions for answers to these exercises can be found at the back of the book, immediately before the index.

10.1 New housing estate

New Housing Estate comprising four blocks of low-rise flats, a block of five-storey flats, and a shops and maisonettes complex. New water and sewerage systems are required, and an existing drain requires diverting. Attractive landscaping, fencing and screening will complete the site.

Produce a programme for the project and resource level the programme for (a) bricklayers, and (b) total labour.

Restraints

Target completion: 82 weeks
Flats A and B to be completed as early as possible.
Shops, maisonettes not required until all other work on site is completed.

Resource demand

Flats, blocks A, B, C and D, per block:	
Excavation, foundation	40 man weeks
Brickwork	120 man weeks
Carpenters	30 man weeks
Finishing trades	70 man weeks
Shops and maisonettes:	
Excavation, foundations	120 man weeks
Concrete frame	150 weeks
Brickwork, blockwork	150 man weeks
Finishing trades	250 man weeks
Five-storey block:	
Excavation, foundations	190 man weeks
Concrete frame	300 man weeks
Brickwork	350 man weeks
Finishing trades	500 man weeks
Site establishment	4 men × 3 weeks
Drain diversion	30 man weeks
Completion of main drain	24 man weeks 300mm diam drain
Drain connections	40 man weeks connections to main

Roads – 1st phase	60 man weeks
Road surfacing	50 man weeks
Paths	75 man weeks
Fences and screens	20 man weeks
Water services	80 man weeks 150mm diam ring main
Landscaping	100 man weeks

10.2 Pipeline

The contract comprises the laying of two lengths of large-diameter pipeline with flexible joints 16 km from pumping station A to reservoir B and 8 km from B to an industrial consumer C. There is a continuous outcrop of rock from chainage 10 to 13 km, and special river, rail, or road crossings are to be constructed at chainages 2, 7, 13, 18, and 22 km.

Isolating valves are to be installed at ends and at 3 km intervals along the entire length of the main, each providing a suitable flange and anchor for test purposes. Water for testing will be supplied free by the promoter following completion of the pumping station during week 33. The reservoir will be commissioned during week 35. The industrial plant will be completed by week 40, but cannot become operational until water is available. Pipes are available at a maximum rate of 1000 m/week (commencing week 1). The contractor is responsible for offloading from suppliers' lorries, storing and stringing out. Each stringing gang can handle 1000 m/week.

Produce a time-location programme for the contract on the assumption that one stringing gang and four separate pipelaying gangs are to be employed.

The average rate of pipelaying per gang may be taken as 300 and 75 m/week/gang in normal ground and rock, respectively. Testing and cleaning each 3 km length will take 2 weeks. A river, rail, or road crossing is estimated to occupy a bridging gang for 4 weeks. Because of access problems, pipelaying and stringing gangs should not be operating concurrently in the same 1 km length. Attention should be paid to manpower resources and continuity of work.

10.3 Industrial project

(a) Construct a precedence diagram as the master programme for an industrial project using the logic and estimated duration given in the following table. Determine the earliest completion date and mark the critical path.

Activity	Duration (months)	Precedence activity
1. Promoter's brief	3	—
2. Feasibility study	18	1
3. Promoter considers report	12	2
4. Land purchase	12	3
5. Site investigation	4	3
6. Design stage I	6	3
7. Design stage II	4	5,6
8. Civil tender documents	3	5,6
9. Specify mechanical plant	3	6
10. Mechanical plant tender	4	9
11. Civil tender	2	7,8
12. Specify electrical plant	3	9
13. Electrical plant tender	5	12
14. Manufacture mechanical plant	18	10
15. Design stage III	4	7,10
16. Design stage IV	4	13,15
17. Manufacture electrical plant	20	13
18. Construction stage I	6	4,11,15
19. Construction stage II	12	16,18
20. Install plant	6	14,16,17,18
21. Test and commission	3	19,20

Estimate the float associated with activity 16.

(b) The tenders for the electrical and mechanical plant have been received and are as follows:

Plant		Period of manufacture (months)	Cost
Electrical A	Activity 17	20	444 000
Electrical B	Activity 17	16	600 000
Mechanical C	Activity 14	18	600 000
Mechanical D	Activity 14	16	700 000

If the promoter's profit is estimated to be £50 000/month from the date of the completion, which two tenders would you recommend?

10.4 Bridge

Using the precedence diagram method you are to produce networks for the construction of the bridge shown in Figure 10.7.

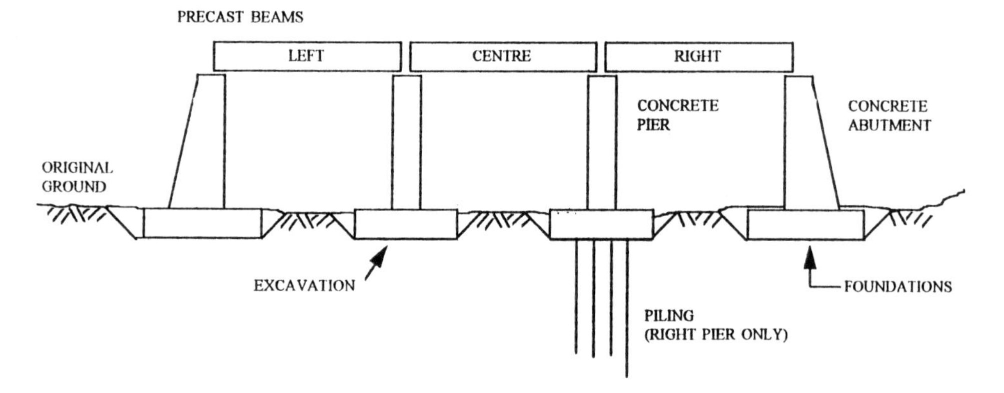

Figure 10.7 Example of bridge construction.

Activity number	Activity description	Duration (weeks)	Resource demand
1	Set up site	1	
2	Excavate left abutment	6	Excavation
3	Excavate left pier	4	Excavation
4	Excavate right pier	4	Excavation
5	Excavate right abutment	6	Excavation
6	Piledriving to right pier	7	
7	Foundations left abutment	8	Concrete
8	Foundations left pier	6	Concrete
9	Foundations right pier	6	Concrete
10	Foundations right abutment	8	Concrete
11	Concrete left abutment	9	Concrete
12	Concrete left pier	7	Concrete
13	Concrete right pier	7	Concrete
14	Concrete right abutment	9	Concrete
15	Place beams left span	4	Crane
16	Place beams centre span	4	Crane
17	Place beams right span	4	Crane
18	Clear site	1	Crane

The exercise is in **three** parts.

(a) Produce a network assuming unlimited resources are available. Evaluate the network, showing the critical path and earliest completion date.
(b) By consideration of the network produced in (a), evaluate the effects of reducing the duration of piledriving to 5 weeks.
(c) As (a) but assume the resources are heavily constrained.
The resources available in total are one excavation team, one concrete team for foundations, one concrete team for abutments and piers and one crane team.

Chapter 11

Project Control Using Earned Value Techniques

One of the most significant developments in project control occurred in 1967 when the United States Department of Defense published a set of cost and schedule control system criteria, known as the C-Spec. These criteria define minimum earned value management control system requirements. Previous management control systems assumed a direct relationship between lapsed time, work performed and incurred costs. This chapter describes how the earned value system analyses each of these components independently, comparing actual data with a baseline plan, set at the beginning of the project.

11.1 Definitions

Earned value analysis (EVA): compares the value of work done with the value of work that should have been done.

Budgeted cost of work scheduled (BCWS): the value of work that should have been done at a given point in time. This takes the work planned to have been done and the budget for each task, indicating the portion of the budget planned to have been used.

Budgeted cost of work performed (BCWP): *the value of the work done at a point* in time. This takes the work that has been done and the budget for each task, indicating what portion of the budget ought to have been used to achieve it.

Actual cost of work performed (ACWP): *the actual cost of the work done.*

Figure 11.1 illustrates a typical S-curve plot comparing budget and actual costs.

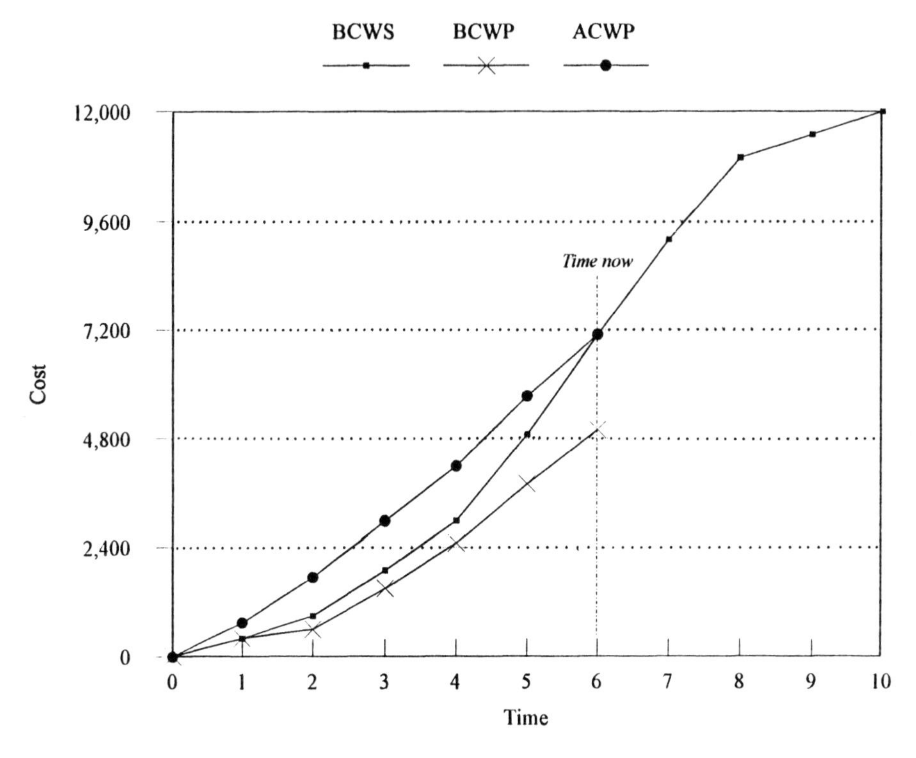

Figure 11.1 Typical S-curve.

Productivity factor: the ratio of the estimated man-hours to the actual man-hours.

Schedule variance (SV): the value of the work done minus the value of the work that should have done (BCWP – BCWS). A negative number implies that work is behind schedule.

Cost variance (CV): the budgeted cost of the work done to date minus the actual cost of the work done to date (BCWP – ACWP). A negative number implies a current budget overrun.

Variance at completion: the budget (baseline) at completion (BAC) minus estimate at completion. A negative value implies that the project is over budget.

Schedule performance index (SPI): (BCWP/BCWS) × 100. Values under 100 indicate that the project is over budget or behind schedule.

Cost performance index (CPI): *(BCWP/ACWP)* × 100. Values under 100 indicate that the project is over budget.

11.2 Theory and development

S-curves examine the progress of the project and forecast expenditure in terms of man-hours or money. This is compared with the actual expenditure as the project progresses, or the value of work done. If percentages are used, the development of useful data from historic records of past projects is simplified, since size, and hence total time and cost is no longer significant. All projects, whatever their size, are plotted against the same parameters and characteristic curves can be more readily seen.

The form of the S-curve is determined by the start date, the end date and the manner in which the value of work done is assessed. Once a consistent approach has been established and the historical data analysed there are three significant variables that need analysing: time, money and the shape of the S (known as the *route*). As the expectation is that the route is fixed, then only two variables are left. The route is as much a target as the final cost. If the movement month to month is compared then the trend can be derived.

If the axes are expressed as percentages, then care is needed in defining what the percentages are based on. If review estimates are made at regular intervals then the value of work done will always be a percentage of the latest estimate. Revision of the estimate automatically revises the route of the S-curve.

The assessment and precise recording of value of work done is crucial to project cost control. This is described by accountants as *work in progress*. For example, design man-hours are usually measured weekly. The hours that have been booked can be evaluated at an average rate per man-hour, that rate being the actual costs in the project. Materials are delivered against a firm order, so normally an order value is available. The establishment of realistic targets is very important if the analysis is to be meaningful. A low cost estimate often leads to a low estimate of the time required to carry out the work. The immediate target is not the final target.

A series of standard S-curves have been developed by companies in the oil and gas industries so that the performance of existing projects can be monitored. These curves have been derived empirically, and when projects within certain categories do not follow the norm, investigations ensue to identify the source of the discrepancy. These curves have been put to a number of uses, including monitoring, reporting, and payment. They are not always used in the pure form: that is, further curves such as productivity factors can be developed so that certain aspects of the project can form the focus of attention.

Developments have been made from using the S-curve simply as a method for controlling the progress of the project to using it proactively in the evaluation of indirect costs associated with project changes introduced by the promoter. This method is known as *impact* or *influence*.

Influence may be applied whenever there is a variation; it is applied to the estimate of additional man-hours taking into account the indirect costs for the whole of the variation, including the additional indirect costs associated with parallel activities. The revised estimated cost of the variation may then be issued to the promoter.

The influence is composed of:

- time lost in stopping and starting current activities in order to make the change;
- special handling to meet a previously scheduled activity;
- revisions to project reports and documents;
- unusual circumstances that could not be foreseen;
- recycle (lost effort on work already produced);
- other costs that may not appear to be related to a particular change.

Influence is incorporated by multiplying the direct cost for the variation by the influence factor and adding this to the direct cost to give the total cost for the change:

$$T = V(1 + IF) \qquad \textbf{(11.1)}$$

where: V = direct cost of variation
T = total cost of variation
IF = influence factor
Cost = cost to the promoter

Any organization wishing to adopt this approach would need to derive standard curves.

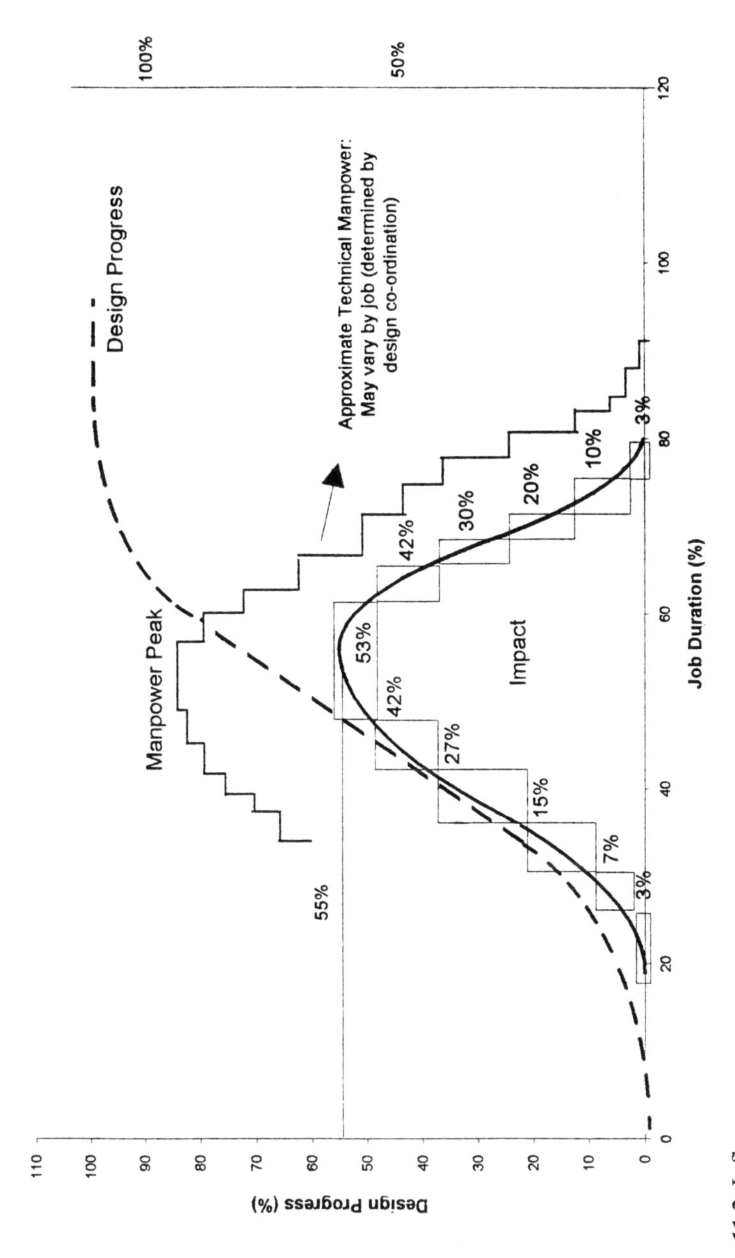

Figure 11.2 Influence curve.

11.3 Relationship of project functions

Monitoring and cost control can be described as identifying what is happening and then responding to it. Cost planning involves forecasting how money will be spent on a project in order to determine whether the project should be sanctioned, and having sufficient money available when required.

Typically there are three major areas of control: *commitments*, *value of work done* and *expenditure*. These are all controlled in relation to their progress over time and may be illustrated diagrammatically, as shown in Figure 11.3, which presents the typical S-curve for the value of work done. Similar S-curves may be developed both for commitments and for expenditure.

Planning are primarily responsible for establishing the time target, both overall and in detail. The primary, although not the only, task of project cost control is to establish the exact position of the project from time period to time period in terms of value of work done, and compare this with the targets for each time period. The finance department will maintain expenditure records, working closely with project cost control.

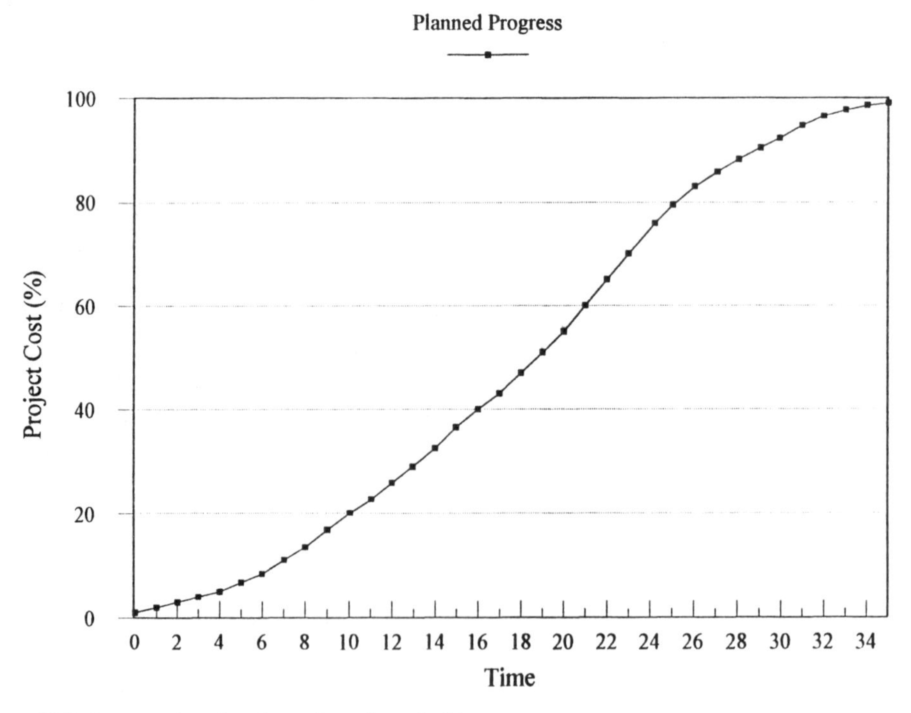

Figure 11.3 S-curve showing the value of work done.

They will be responsible for maintaining a record of the commitments, as they have to ensure that payments are within the approved limits.

Project control can comment on the validity of planning work, by comparing planned and actual progress via the value of work done. The commitment record, if properly maintained, provides a ceiling at any point during the life of the project.

11.4 Value of work done control

Value of work done is not expenditure, although it eventually equates to expenditure. It is often considered as the work in progress but is, most of the time, greater than the expenditure. Figure 11.4 illustrates the relationship between value of work done and expenditure as recorded over the life of a project.

There is normally a considerable time-lag between having a project ready for start-up and the final payment of invoices and retention when the project is completed in financial terms. The value of work done can be summarized as design and head office costs plus the value of material delivered to site and the work done at site.

The techniques for an approximation of value of work done from month to month can be related to three major areas: *head office*, *material deliveries to site* and *erection*.

Head Office

This deals with design, procurement and project supervision activities. The work done can be measured in man-hours and is usually recorded on a weekly basis. Hours booked are valued at an average man-hour rate and the value of work done then assessed. The simple way to do this is on a cumulative basis. If the latest cumulative booked cost comes perhaps three weeks later, then that is compared with the cumulative cost including the last month's estimate; a correction can be made resulting in the approximation being only one month behind the actual cost at that time.

Materials delivery to site

Materials are normally delivered to site against firm orders so an order value is available. As materials arrive on site the material receipt note can be valued, using information on the order, and a progressive total maintained. As the value shown on the order may not be the absolute

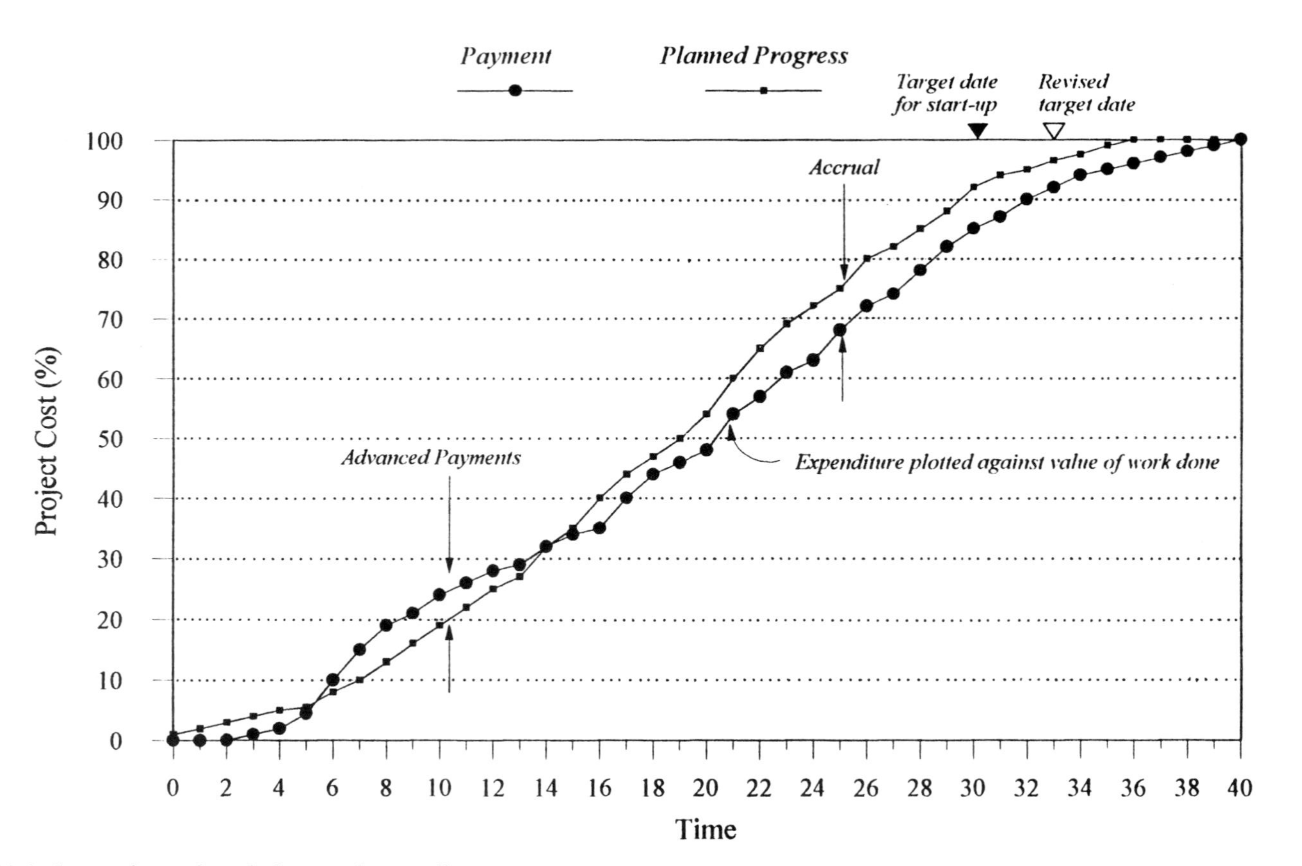

Figure 11.4 Comparison of work done and expenditure.

invoice value, for reasons such as discounts, freight charges, insurance or escalation, this approach gives a nominal and close approximation to the value.

For bulk materials it can be very time-consuming to value all items on a receipt note from the itemized prices on an order, and it is often permissible to use a weighbridge weight multiplied by an average price per kilogram calculated from the order, or orders.

Erection

The approach to erection contracts is very similar, but this time the value is erection man-hours. Erection man-hours, exclusive of site supervision, for all contracts on site, should be recorded from week to week. Progress on site by various erection contractors is measured by teams of schedulers or measurement engineers in accordance with certain standard methods of measurement to enable progress payments to be made. Such evaluations often run late, so erection man-hours worked are often valued on the basis of standards multiplied by an average rate per hour, applicable to the erection activity.

11.5 Earned value analysis techniques

Earned value analysis (EVA) is often presented in the form of progress, productivity or S-curve diagrams. Many of the current reporting procedures were developed by the US Corps of Engineers. EVA has been adopted by the oil, manufacturing, gas and process industries, where man-hours and material deliveries to site are used to monitor projects from inception to completion.

Actual/estimated man-hours are made available to determine the progress and productivity factors at any stage of the project. The productivity factors are used by both promoter and contractor organizations to monitor the progress of a project and forecast the outcome.

Productivity is the ratio between output and input and provides a measure of efficiency. Ideally, productivity is always unity, with both the output and the input measured in the same units. In the oil and gas industries man-hours are used as the common unit to determine productivity. Productivity factors are used to monitor variance and trends for individual activities.

In order to establish a trend, actual progress must be measured and compared with forecasted progress, that being the baseline S-curve. As the forecasted progress depends upon the end target, revised targets will influence the progress required to be made each month. Once it becomes evident that a work package is going to cost more, or less, than the original (or earlier) estimate, then targets should be revised, the potential influence on monthly progress evaluated, and a new target computed.

Table 11.1 illustrates typical 'progress of value of work done' expressed both in man-hours and completed items, common to manufacturing industries. The revised number of man-hours is forecast against the actual work completed and is based on progress and productivity. Often this information is displayed on a screen to manufacturing operatives as a guide to production rates and the time to complete. If a bottleneck occurs at any point in the production line the effect on the costs, time and productivity of the completed item can be identified. In the case of production being carried out at a number of locations with final assembly at one specific point it is imperative that any deviation from the activity S-curve is reported immediately and acted upon.

Table 11.1 Progress of value of work done.

	Man-hours	Units
To date		
Estimated	360 000	4750
Actual	295 000	4900
Total		
Estimated	500 000	6000
Revised	440 000	6000

In many manufacturing industries, especially automobile assembly lines, man-hours are allocated for reworking items to meet quality and standards. Mercedes Benz for example, reportedly spend up to 30% of the total man-hours allotted to reworking to meet defined quality and standards. This *manhour float* is often the determinant for price setting, to the customer, for each model. The hours expended from the float of man-hours is then used to forecast the final costs, times and margins. In many cases reworking is analogous to commissioning, and as such, may not be allotted targeted man-hours in the project estimate.

11.6 Applications of EVA

The information required from analysis of the curves will vary depending on its end use. At project level the aim will be to identify any areas where the project is underachieving, in order that action can be taken to improve the performance of the problematic resource. This can be done by examining the BCWP and ACWP curves. If the curves are showing cause for concern then productivity curves can be plotted and examined in greater detail. Work packages or sets of work packages can be examined and the productivity derived.

A project manager can adopt this system through analysis of completed time sheets. It is also important to note the percentage complete status of a task at the recording date, as described earlier. This is where a proactive role is required by the project management team in gleaning information from the various disciplines. The team leader should progress the work of his or her section, with the planners taking a recording and reporting function. If progress is to be expedited then the project team must actively pursue information, check its accuracy and take action when low productivity becomes apparent.

Optimum workforce requirements will become apparent as data for specific projects are collated. Productivity increases or decreases will permit management decisions based on actual measurement rather than optimistic or pessimistic forecasts based on 'gut feeling'. The team leaders of various disciplines must be aware of the significance of the information they are reporting and why it is being recorded: if they do not realize that it will be used for control and not just for monitoring then they may not give it a high priority, resulting in erroneous data.

Cost codes for both direct and indirect cost will need to be considered on the basis of compatible codes allocated to different work packages. In a number of major process organizations in the oil and gas industries, for example, the number of multidisciplinary functions and development stages of a project has led to confusion of the cost codes, resulting in inaccurate data and delays in transmitting data for analysis. Ideal cost-coding systems are those that allow the system to be used at all stages in the development of a project.

One very important aspect of coding and recording man-hours is the comparison of man-hours expended against a code or codes and the remaining man-hours expended as overheads. If for example a design engineer is allocated 40 hours per week on say three projects and only 32 hours are recorded, then the remaining 8 hours must be reconciled somewhere within the organization. Unfortunately, the remaining man-hours are often distributed between the project codes. In cost-plus contracts it is standard procedure to book additional man-hours against the contract code.

To achieve the introduction of EVA a basic requirement is the collection and processing of data related to existing and past projects.

The importance of developing standard curves cannot be overemphasized. If there are no standard curves then it is difficult for the project manager to set realistic targets.

These data form the basis for estimating and allotting man-hours to each activity or work package. The effort required at project level to undertake EVA is such that it is not recommended for small projects. Below a certain size of project, the effort required to gather and process data on the progress of small value items may be greater than the value of the package of work. The proportion of the inaccurate part of the assessment of the work left will also be greater, invalidating the approach.

If the decision is made to adopt EVA for a given project then the management team must be dedicated to fully implementing the system. Taking action on inaccurate data can be worse than taking no action at all. A standard reporting technique should be developed for all projects so that a given productivity rate means something.

11.7 Examples of EVA

Table 11.2 illustrates the historical data for a typical six-package project. Estimated man-hours are often based on historical data from completed projects against which the actual man-hours expended to complete each work package and the overall project are plotted.

Table 11.2 Work packages, man-hours and productivity factors.

Work package	Actual man-hours	Estimated man-hours	Productivity factor
Feasibility study	550	450	0.82
Design	3000	2500	0.83
Prototype	1500	1700	1.13
Manufacture	2300	2200	0.96
Erect	900	600	0.66
Commission	300	300	1.00

Figure 11.5 illustrates the S-curve for the prototype work package shown in Table 11.2. The S-curve is prepared on the basis of estimated man-hours. The actual man-hours expended are then plotted on a regular basis to determine the variance in man-hours and productivity factors and to forecast the work package trend.

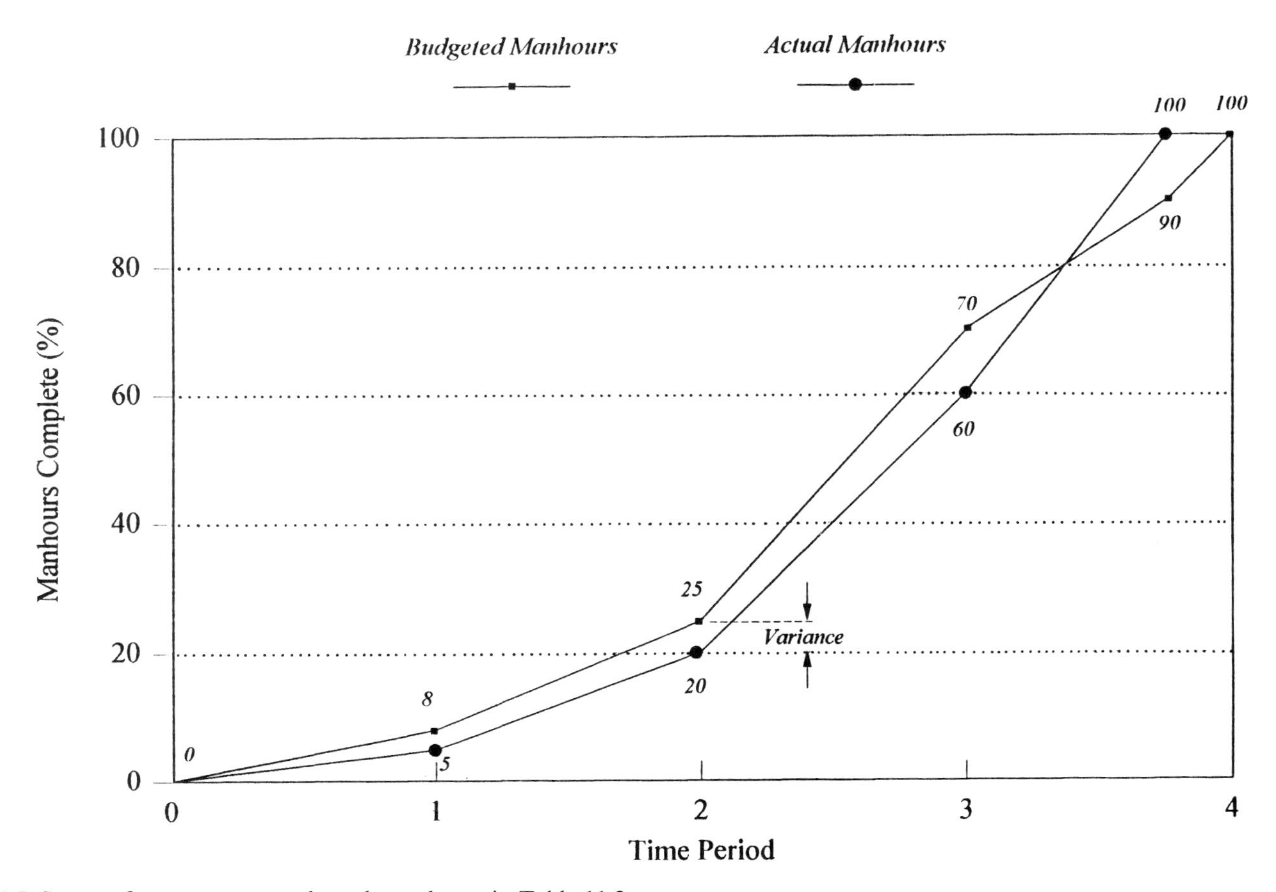

Figure 11.5 S-curve for prototype work package shown in Table 11.2.

On a number of civil engineering projects EVA has been used primarily to monitor the construction of individual units and as a reporting tool to both construction management and the promoter. In its simplest form, bar charts are prepared from which histograms of man-hour estimates for each construction activity are generated. The cumulative man-hours are plotted in the form of an S-curve and progress is monitored as man-hours are expended. Plant usage is often added to man-hour schedules based on weekly costs, usually plant hire rates, for specific types of plant/activity usage. The type of plant and estimated usage often form the basis of plant allocation for a number of projects. Clearly, if the productivity of one particular work package falls below the estimate illustrated by the S-curve then additional plant time and usage will be required to reverse the trend.

The diagrammatic presentation facilitates the comparison of actual man-hours expended in construction with the estimated man-hours. If, for example, in a fixed-price construction project the foundation works of a particular unit has a programme duration of 4 weeks and 480 hours and is completed in three weeks utilizing only 320 hours, then the saving in man-hours can be utilized on another construction unit. The saving in man-hours and plant usage and the reduction in duration can then be plotted as actuals against the estimated S-curve. The 'productivity factor' for this completed activity would be 1.5.

If, however, the man-hours expended are 480 hours after 4 weeks with only 50% of the activity completed then completion would require another 4 weeks' duration and an additional 480 hours based on the previous productivity factor. This often results in manpower being reallocated to an activity falling below the estimated productivity factor. Also, the requirement of plant to complete the work package will often result in delays to other work packages, as it is normally very difficult to acquire plant at short notice.

A major problem area is the accuracy of reports of percentages of work completed. As a result of this, activities or work packages are often reported as being completed when in fact they are not. This is often referred to as the 'persistent 99% complete syndrome', which results in the saying that 99% of tasks in 99% of projects are 99% complete for 99% of the time. It should also be noted that the method of appraising work completed should be timely and consistent for each section of work and always kept up to date. In many civil engineering contracts managers prefer to rely on the 'remaining duration' as a good measure of progress for a work package, as this does not assume anything about the accuracy of the estimate or when the work package started. This approach also ensures that the manager is aware of the remaining duration to meet any contractual milestones, and able to act accordingly.

EVA is also used in the manufacturing and production industries to monitor production rates. In these industries, work is normally performed in controlled environments where it is a simple task to measure the physical work done and materials used, especially where the delivery of an item is paramount to the success of a project. By monitoring, often on a daily basis, the number of man-hours both estimated and expended on a number of activities, delivery dates can be identified at an early stage of the project. Cumulative man-hours expended for each activity are compared with the estimated target of man-hours and the completed items, which permits adjustments to the overall target over the manufacturing period. Changes in the base rates/hour for each activity can easily be amended to identify the cost to completion. In production line industries man-hours expended can be collated on a daily basis through time sheets and completion verified, and hence accurately reported on the basis of a completed item.

Many of the US companies involved in the oil and gas industries in the Gulf of Mexico introduced EVA to major projects carried out in the Middle East. In most cases these projects were undertaken on a cost-plus basis. Data, in the form of man-hours relating to specific tasks, were analysed and then utilized to provide 'productivity factors' relating to work carried out on projects in the USA. To compensate for the required time to learn specific tasks, productivity factors were factored to take into account the additional times required.

For example, the work package of cold insulation of pipework may have a US base rate of one man-hour to insulate 1 m of 200 mm diameter pipe. To take account of the possible learning period for this activity, often performed by third country nationals (TCN), a factor of say 1.2 is used. The S-curve is then factored by 1.2: this factor is called the US Gulf factor (USGF). If the total number of estimated man-hours is say 200 000 hours then the factored total will be 240 000 hours.

The man-hours can then be scheduled for this work package against time, as illustrated in Table 11.3.

Clearly the overall 'productivity factor' for this work package is 0.96; however, productivity over the 9 month duration has varied from 1.33 to 0.5. This method of reporting and planning at timely intervals provides management with a clear view of progress and productivity and provides the basis for management decisions. In this particular example of a cost-plus project the additional man-hours expended to meet the contract programme are paid for by the promoter.

It is very important that work packages utilized in EVA are defined correctly, in the disciplines to be adopted, in the allocation of man-hours,

Table 11.3 Estimated and actual man-hours and productivity factors.

Month	Estimated man-hours	Actual man-hours	Productivity factor
1	10 000	18 000	0.55
2	40 000	50 000	0.80
3	60 000	70 000	0.86
4	50 000	48 000	1.04
5	40 000	30 000	1.33
6	20 000	15 000	1.33
7	10 000	8 000	1.25
8	9 000	10 000	0.90
9	1 000	2 000	0.50
	240 000	251 000	0.96

and in the interdependence on other work packages. The overall project is determined by the worst work package. In a project consisting of say 17 work packages, productivity of 16 of the packages may be above that estimated. In a number of projects contractors have superimposed individual S-curves to provide a total contract S-curve, normally as the basis for reporting to the promoter. This often results in a false picture of the project's progress, as the total contract S-curve implies that the project is ahead of schedule.

In one example, schedulers on one particular work package reported that productivity was as estimated; however, the man-hours recorded against the actual work completed were optimistic. This resulted in the project being reported as being ahead of time and below budget. When an audit of the work package was performed it was found that unless this critical work package was brought back on programme the remaining 16 work packages could not be completed.

An additional 30 000 man-hours had to be allotted to the work package, which resulted in an overall productivity factor of 0.33. Clearly the accuracy of reporting actuals and progress is a prime function of EVA. Should effective reporting and updating have been carried out then at some time during the activity negative progress would have alerted management of the apparent problems.

A further problem on this particular project was monitoring the receipt of materials delivered to site. The numerous cost codes, often with more than one cost code being applied to preassembled units, resulted in delays in progressing invoices. In some cases a three-month lag occurred between receipt of materials and the processing of invoices.

To ensure that the same mistakes did not occur again the promoter instructed the contractor to change the scales on each work package S-curve. Clearly when a project has an estimated manhour expenditure of 11 000 000 hours a slight deviation on a graph does not fully represent the impact on the project as a whole. Over the remaining project duration each work package was expressed in total histogram form with 'productivity factors', time to complete and trend analysis illustrated for presentation to the promoter.

11.8 Summary

Standard procedures need to be prepared as the EVA system is tested and the requirement of each discipline addressed. As with many reporting and scheduling systems it is essential that the organization use and develop the system to suit its needs, and not to create mountains of irrelevant information.

EVA is primarily a system of approximation, the accuracy of which depends on the time and costs prepared in the estimate compared with the actual time and costs as work progresses. The accuracy of the estimated data and actual data and the time intervals for auditing are paramount to its successful application.

Further reading

Bower, D., Thompson, P.A., McGowan, P.H. and Horner, R.M.W. (1993) Integrating project time with cost and price data. In: *Information Technology for Engineering Construction* (ed. B.H.V. Topping), Civ-Comp 93, pp. 41–49. Civil-Comp Press, Edinburgh.

Kharbanda, O.P. and Stallworthy, E.A. (1991) *Cost Control*, Institution of Chemical Engineers, Rugby.

Reiss, G. (1991) *Project Management Demystified.* E. & F.N. Spon, London.

Chapter 12
Contract Strategy

Contracts are used to procure people, plant, equipment, materials and services. Contracts are therefore fundamental to the management of almost all engineering projects. The type of contract should be selected only after consideration of the nature of the parties to the project, the project objectives and the equitable allocation of duties, responsibilities and risk. This chapter outlines the main components of the process used to determine how the project will be procured, usually referred to as the *contract strategy*.

12.1 Contractual issues

- When is an agreement a contract?
- Is a purchase order a contract?
- What are conditions of contract?
- Why are there different types of contract and various terms of payment, and how do they motivate contractors and others?
- What is needed for an internal equivalent of a contract between the engineering and the production, purchasing or construction departments within an organization?

Experience should provide some knowledge and ideas for answering these questions, but it can be incomplete or misleading. This chapter is not meant to be a substitute for the expertise and thinking needed in the drafting and the administration of contracts. It is an introduction to the relationship between legal means and engineering ends, to be followed up by further reading to suit individual jobs and needs.

Most contracts for engineering and construction work include a list of definitions of some of the words used in them. This is wise and helpful, but unfortunately the definitions vary from contract to contract and so there are no standard meanings used consistently even within the UK. As far as possible the most typical meanings are used here.

12.2 Contract planning

In the planning of a contract the promoter needs to consider carefully the motive in employing a contractor. Promoters generally employ contractors for one or more of the following reasons:

- to utilize the skills and expertise of contractors, managers, engineers, craftsmen, buyers, etc. for the limited duration of a project;
- to have the benefit of contractors' special resources such as licensed processes, unique plant, materials in stock;
- to get work started quicker than would be possible by recruiting and training direct employees;
- to get contractors to take some of the cost risks of a project, usually the risks of planning to use people, plant, materials and sub-contractors economically;
- to get contractors to obtain the financial resources for a project, and perhaps also for operating it;
- to be free as promoters to concentrate on the objectives of projects, their subsequent use, and other interests;
- to encourage the development of successful contractors.

One of these reasons must have priority in deciding to employ a contractor, and the others given lesser or no importance. Whichever has priority should govern the decisions on the number, scope, type and terms of contracts.

There is a basic conflict between the risk and motive provisions in most contracts. The contract includes provisions for allocating risk between the parties and identifying the implications for time and cost.

Number of contracts

The number and sequence of contracts for goods and services can vary greatly from project to project. Often a consultant or design contractor is employed just for the feasibility stage of a proposed project, and a project management consultant to advise on risks and contract strategy. For implementation, it may be appropriate for a promoter to employ a contractor for only a part of a project and one or more others for the remainder. Or a project might be so large that more than one contractor is appropriate to share the risks. For the equipment required for a new factory one contractor might be employed to install equipment supplied by others.

A series of contractors can be employed in turn for construction work: for instance, one for demolition work, another for new foundations, the superstructure and building work, and others for designing and supplying equipment, installing it, completing systems and services, testing and commissioning for replacing part of a factory, each under different terms of contract. For a building project, different contractors could be employed separately for the structural, finishing and services work, instead of one main contractor; these specialist tasks could be sublet, or a joint venture or consortium approach could be adopted.

Model forms

Most project managers make use of model or standard forms of contract. These are conditions of contract that have been prepared for general use in a particular industrial sector by an appropriate or representative authority. For example, in the civil engineering sector the Institution of Civil Engineers Conditions of Contract are drafted by the Institution of Civil Engineers, the Association of Consulting Engineers and the Federation of Civil Engineering Contractors. The advantages of adopting a tried and tested form of contract are seen to be: wide acceptability, consistency of use, familiarity in use and legal case law precedents. Model forms are discussed further in section 12.8.

12.3 Contract strategy

The development of a contract strategy for any project should be based on a thorough assessment of the choices available for the implementation and management of design and construction.

The main topics for consideration can be subdivided as follows:

- project objectives
- the organization system for design and implementation
- risk allocation
- the terms of payment
- the conditions of contract
- the tendering procedure.

The first step is to identify clearly the project objectives and the main constraints on implementation.

12.4 Project objectives

It is essential that the project manager ensures that the promoter defines the project objectives clearly, as the likelihood of a successful project is greatly improved when all the managers of the design, construction and supporting groups are fully informed and committed to these objectives. The project objectives should also be communicated to the other parties involved in project implementation.

The normal project objectives are concerned with cost, time and quality (Figure 12.1). These are interrelated and may often conflict:

- minimum time, timely completion, early start to construction;
- minimum cost, fixed budget;
- high or appropriate quality.

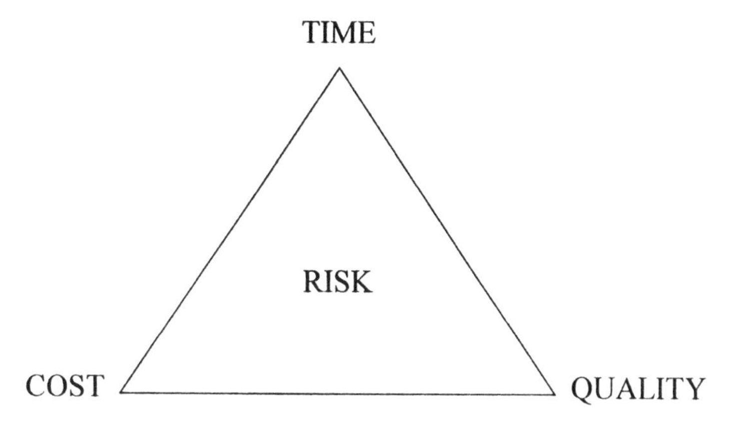

Figure 12.1 Project objectives.

On an increasing number of projects, and especially those in developing countries, other objectives can be dominant; for example:

- the management and control of risk;
- training of the promoter's staff;
- retention of construction plant by the promoter;
- promoter involvement in project and construction management;
- cooperation of construction and implementation contractors in design;
- substitution of labour-based construction;
- establishment of a maintenance force.

Project management should rank the objectives and seek implementation policies that optimize their achievement.

12.5 Organizational choices

The choice of appropriate organization for design, construction, and commissioning where relevant, should consider:

- the ability to meet project objectives;
- the resources and services offered;
- the resources and expertise that the promoter is able to commit to the project;
- the balance between management, design and construction.

Package deal (turnkey, design and supply)

The simplest arrangement contractually is that in which one contractor is responsible for all decisions until hand-over of the completed work to the promoter. Payment is generally on a lump sum basis, although this is often broken down into elements or phases of work. One contractor is responsible for engineering, procurement, construction, installation of equipment, testing and commissioning, although parts of the work may be subcontracted to specialists; and may be responsible for its financing, obtaining public approvals, purchasing process materials or other functions until hand-over of the completed work to the promoter.

The main strengths of the package deal contract are as follows.

- A firm price can be obtained at an early stage in the project provided the promoter's requirements are known.
- Time can be shortened from design/construction overlap.
- The promoter deals with only one organization for both design and construction.
- Design should be tailored to give the most efficient construction.
- The cause of defects cannot be a matter of dispute.

There are some weaknesses, which primarily relate to the role of the promoter in the project, including the following.

- The design and build firm may not always act in the promoter's best interests.
- The promoter must commit to the whole package at an early stage.
- The promoter is in a relatively weak position to negotiate change.
- The extent of competition is likely to be reduced.
- In-house practices of the promoter may constrain the design and build firm.

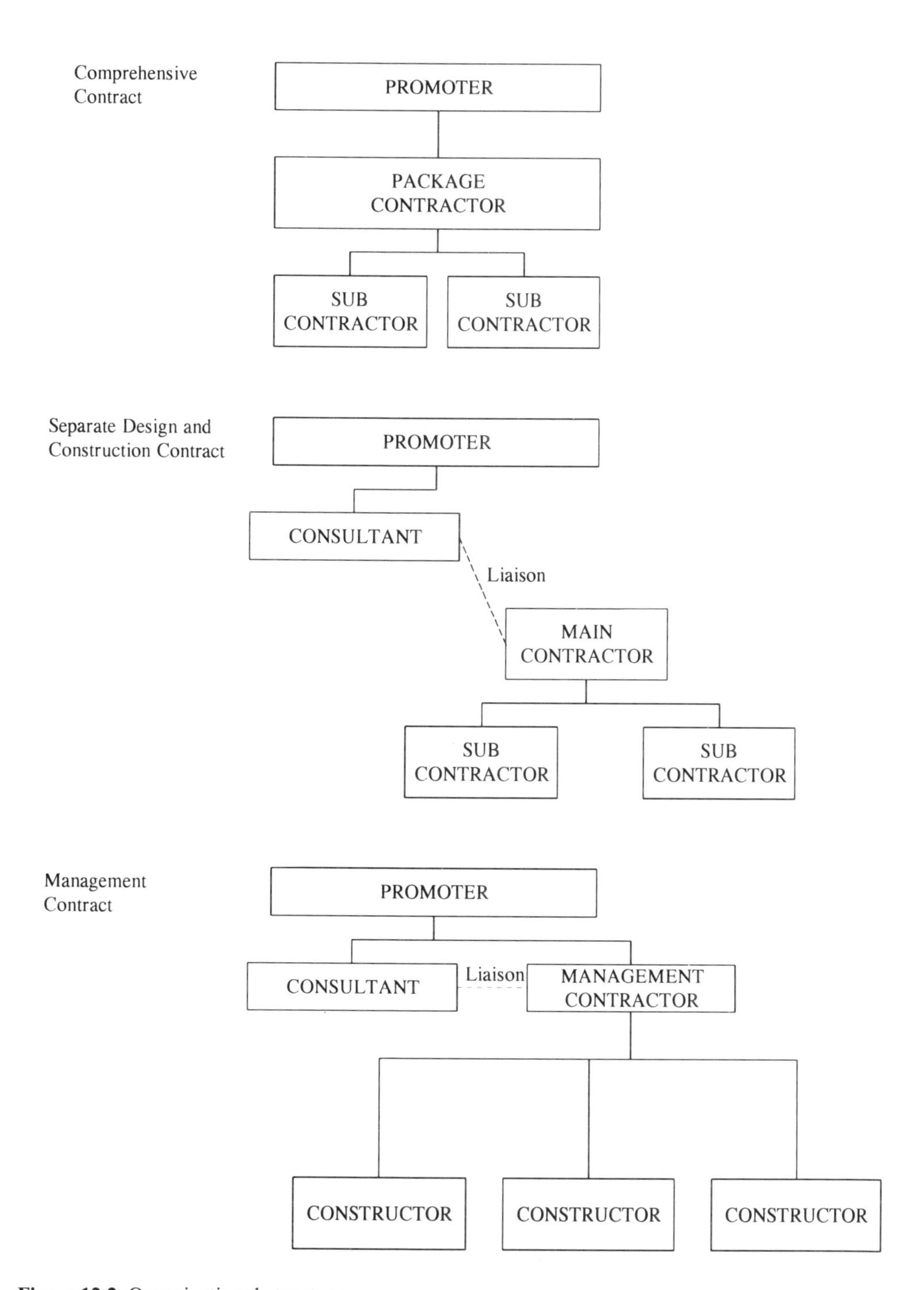

Figure 12.2 Organizational structure.

The package deal contract can be used when the building is of a standard or repetitive nature – for example, houses, warehouses, office blocks – and it can also be used where contractors offer specialist design/construction expertise for particular types of project. It can satisfy a need for an early start to a project. When the promoter and his advisers have insufficient specialist management resources for the project, or if the promoter wishes to place the work with a single organization, this form of contract is effective.

Build-own-operate-transfer contracts

The concept of build-own-operate-transfer (BOOT) can be defined as a major start-up business for which private organizations undertake to build, own and operate a project that would normally be undertaken by government or a private promoter organization. They then return the ownership of the facility to the government or private promoter organization after a fixed concession period. These contracts are described in more detail in Chapter 15.

Separation of design and implementation

One common organizational structure is a series of two or more contracts for the successive stages of work for a project. A common example is the employment of a consultant or design contractor for the design of a project, and then a main contractor for its implementation. In the UK this is known as the *conventional* or *traditional* contract arrangement, particularly for building and civil engineering projects.

Design and supervision of construction and implementation is carried out by consulting engineers or architects or design contractors. Construction is undertaken by a contractor, usually under unit rate or occasionally a lump sum contract. Management responsibilities are divided.

Management contracting

Management contracting is an arrangement in which the promoter appoints an external organization to manage and coordinate the design and construction phases of a project. The management organization does not normally execute any of the permanent works, which are packaged into a number of discrete contracts, but may provide specified common user and service facilities.

When management contracting is used the promoter is creating a contractual and organizational system that is significantly different from the conventional approach. The management contractor becomes a member of the promoter's team and the promoter's involvement in the project tends to increase.

Payment for the management contractor's staff is reimbursable plus a fee, but the engineering contracts, let by the management contractor, will usually be lump sum or unit rate. The management contractor is appointed early in the project life and has considerable design input. Designers and contractors are employed in the normal way. Another benefit is that the work can be divided into two or more parallel simultaneous contracts, offering reduced durations and risk sharing. Separate contracts for independent parts of the construction of a project divide a large amount of work amongst several contractors.

Some contracts would be consecutive but it is possible to award two or more parallel simultaneous contracts. Examples in industrial construction and maintenance are separate contracts to employ different contractors' experience on different types of work.

Time saving is one of the main advantages of this form of contract:

- Time can be saved by extensive overlapping of design and engineering. By saving time, the impact of cost inflation can be reduced.
- Cost savings can also arise from better control of design change, improved buildability, improved planning of design and engineering into packages for phased tendering, and keener prices due to greater competition and better packaging of the work to suit the contractor's capabilities.
- Further time saving arises from the management contractor's special experience, for example overcoming shortages of critical materials, and delays arising from a scarcity of particular trades.
- It provides flexibility for design change during the construction arising from the promoter, to improve the fitness for his needs, and from the contractor, to improve buildability.
- Discipline can be imposed on the promoter's decision-making procedures and on the management of design, in particular to ensure proper coordination of the availability of design information with the sequence of work, particularly on multidisciplinary projects.

There are also some potential drawbacks with management contracting:

- The promoter may be exposed to greater risk from the contractors than in a conventional arrangement.

- There are risks due to the absence of an overall tender price for the complete works at the start of work.
- There is a tendency to produce additional administration and some duplication of staff on supervision.

Management contracts can be used when there is a need for:

- early start to the project for political reasons, budgetary or procurement policy.
- early completion of the project but design cannot be completed prior to construction. This circumstance requires good planning and control of the design/construction overlap and careful packaging of construction contracts.
- innovative and high technology projects when it is likely that design change will occur throughout.
- organizational complexity. Typically this may arise from the need to manage and coordinate a considerable number of contractors and contractual interfaces and possibly several design organizations. It is useful when the promoter and his advisers have insufficient specialist management resources for the project.

Process and offshore engineering

In the process plant industry rather different terminology is used from building and civil engineering. Most contractors in that industry offer engineering, procurement and construction and may be employed as EPC contractors, turnkey contractors or construction only. The essential difference between EPC and turnkey is that in the former the various stages are clearly separated and may be paid for in different ways: for example, engineering and procurement may be reimbursable and construction paid for by admeasurement. There is much greater opportunity for promoter involvement than with a turnkey contract.

Project services and management services contracts have been developed to deal with very large and complex projects such as North Sea oilfield developments. These operate in a similar way to management contracts in building and civil engineering but with some significant organizational differences. For example:

- very large numbers of the staff are involved, and 'management of management' becomes crucial to the success of the project;
- the project services contractor is formally integrated into the promoter management structure.

Direct labour

Instead of employing an external organization, a promoter may have equipment made or installed or some projects constructed by an in-house maintenance or construction department, known in the UK as *direct labour* or *direct works*. If so, in all but the smallest organizations the design decisions and the consequent manufacturing, installation and construction work are usually the responsibility of different departments. To make their separate responsibilities clear, the order instructing work to be done to the design may be in effect the equivalent of a contract that specifies the scope, standards and price of the work as if the departments were separate companies.

This arrangement should have the advantages of making clear the responsibilities for project costs due to design and due to the consequent work. Except that disputes between the departments would be managerial rather than legal problems, these internal 'contracts' can be similar to commercial agreements between organizations. Contractual principles and these notes should therefore be applied to them.

As described in Chapter 15, if a contractor promotes as well as carries out a project he may need to separate these two roles because different expertise and responsibilities are involved in deciding whether to proceed with the project and then how to do it. Separation of these responsibilities may also be required because others are participating in financing the project. For all such projects except small ones an internal contract may therefore be appropriate to define responsibilities and liabilities.

12.6 Risk allocation

A prime function of the contract is to allocate risk. The identification and consideration of risk is a logical way to develop the organizational and contractual policies for any project. Some of these uncertainties will remain whatever type of contract is adopted, and the tender must include a contingency sum for them.

Some risks can be due to a promoter's uncertainty at the start in specifying what is wanted. If the contract is to be for the design and supply of equipment that is to be part of a system which will be defined by discussions during the contract, or if the initial prediction of the purpose of a project may have to be varied during the work in order to meet changes in forecasts of the demand for the goods or services that it is to produce when completed, then uncertainty is inherent in the project.

Promoters rarely invite tenders on the basis that an organization is to be committed to complete the work regardless of risks; contractors would have to cover themselves by high prices in excess of the most probable direct and indirect costs that they might incur. Governments and other promoters in countries with less engineering expertise at times ask bidders to carry more risk, but the trend in industrialized countries since early in this century has been that a risk should be the responsibility of whichever party is best able to manage it to suit the objectives of a project.

All parties to a project are at risk to some extent whatever the contracts between them: for instance, that work may be frustrated by forces beyond their control. If so, the time lost and all or some of their consequent costs may not be recoverable. The division of risks between the parties is usually established in the conditions of contract specified by promoters when inviting bids.

12.7 Terms of payment

Price-based: lump sum and unit rate (admeasurement) prices or rates are submitted by the contractor in his tender.

Cost-based: cost-reimbursable and target cost. The actual costs incurred by the contractor are reimbursed, together with a fee for overheads and profit.

Lump sum

This type of contract is based on a single tendered price for the whole works. The words 'lump sum' are used in engineering and construction to mean that a contractor is paid on completing a major stage of work or milestone: for instance, on handing over a section of a project. In practice 'lump sum' is also used to mean that the amount to be paid is fixed, based on the bid price but perhaps subject to change for escalation.

The words 'firm' and 'fixed' usually mean that the bid price is to be the final contract price because it will not be subject to escalation. They may also mean that there is no term in the contract for the promoter to order variations. Like other words used in contract management, 'fixed price' has no fixed meaning and 'firm price' no firm meaning. What matters in each contract are the terms of payment in that contract and what the governing law permits.

Payment to a contractor can be in stages, in a series of lump sums each paid upon his achieving a *milestone* – meaning a defined stage of progress. Use of the word 'milestones' usually means that payment is based upon progress in completing what the promoter wants. Payment based upon achieving defined percentages of a contractor's programme of activities is also known as a *planned payment* scheme.

Although this type of contract is relatively inflexible, it is possible to include a 'changes' article which provides a right to make changes within the scope of the works, coupled with the negotiation of price changes.

The advantages of using a lump sum contract are as follows.

- There is a high degree of certainty about the final price.
- Contract administration is easy, provided there is no or little change.
- It facilitates keen pricing.
- The promoter's management resources are freed for other projects.

The disadvantages of using a lump sum contract are as follows.

- It is unsuitable when change is expected.
- There is a possibility that the low bidder may find he is in a loss-making situation, especially where considerable risk has been placed with the contractor. This may lead to cost cutting, trivial claims and, in the extreme, bankruptcy.
- The promoter and design organization have minimal opportunity for involvement in the management of construction.

A lump sum contract can be used to provide an incentive for the contractor to perform, when design is complete at tender and little or no change or risk is envisaged. It can be adopted if the promoter wishes to minimize the resources involved in contract administration and might want to place all or most of the risks with the contractor. This type of contract is rarely used for main civil engineering contracts; however, it is more common on process plants.

Unit rate (admeasurement)

This type of contract is based on *bills of quantities* (BoQ) or *schedules of rates*. Items of work are specified with quantities. Contractors then tender rates against each item. Payment is usually monthly and is derived from measuring quantities of completed work and valuing at rates in the tender, or new rates negotiated from tender rates.

Unit rate terms of payment can provide a basis for paying a contractor in proportion to the amount of work completed and in proportion to the final quantities required by the promoter if different from the amount predicted at the time of agreeing a contract.

For UK building and infrastructure projects the predicted amounts are usually stated in a BoQ, which lists each item of work to be done for the promoter under a contract: for instance, a quantity of concrete to a quality defined in an accompanying specification. An equivalent in some industrial contracts is called a *schedule of measured work*. In these contracts the total price is therefore based on fixed rates but changes if the quantities change.

In some contracts what is called the *schedule of rates* is very similar to a bill of quantities in form and purpose, as contractors when bidding are asked to state rates per unit of items on the basis of indications of possible total quantities in a defined period or within a limit of say $\pm 15\%$ variation of these quantities. In other cases the rates are to be the basis of payment for any quantity of an item which is ordered at any time.

The principal strengths of the admeasurement contract are as follows.

- It is a well-understood, widely used type of contract.
- There is some flexibility for design change.
- There is some overlap of design with construction.
- There is good competition at tender.
- The tender total gives a good indication of final price where the likelihood of change, disruption and risk is low.

The principal weaknesses of the admeasurement contract are as follows.

- Claims resolution is difficult; the contract is quantity based and adversarial.
- There are limits to flexibility; new items of work are difficult to price.
- There are limits to the involvement of the promoter in management.
- The final price may not be determined until long after the works are complete, especially when considerable change and disruption has occurred and major risks have materialized.

The admeasurement contract can be used with the separate organizational structure. It requires that the design is complete but can accommodate changes in quantity. It is used on many public sector civil engineering projects like roads and bridges where little or no change to programme is expected and the level of risk is low and quantifiable.

Admeasurement contracts are sometimes used on high-risk contracts or where considerable change and disruption are expected but the promoter's procedures and regulations prevent the use of a cost-based contract. In these circumstances promoters should proceed with caution and note the deficiencies of traditional BoQ in evaluating extensive change and particularly disruption to programme.

Cost reimbursable

This type of contract is based on payment of actual cost plus a specified fee for overheads and profit. The contractor's cost accounts are open to the promoter (*openbook accounting*). Payments may be monthly in advance, in arrears, or from an imprest account.

The simplest form of cost-reimbursable contract is one under which a promoter pays ('reimburses') all a contractor's actual costs of all his employees on the contract ('payroll burden') and of materials, equipment and payments to subcontractors, plus usually a fixed sum or percentage for financing, overheads and profit. More complicated is a contract under which the costs-plus of achieving all satisfactory or acceptable work are reimbursed, but none or only some of the costs of rejected work.

Contractors working under this low-risk arrangement should not expect to be rewarded at the same rate of profit as under price-based contracts, because the risk burden is less.

Examples are in contracts for novel design, development, installation of pipework and cables, repair, the commissioning of industrial projects, demolition, site clearing and emergency work; where the scope, timing or conditions of work are uncertain. Cost reimbursement is therefore the basis of payment in many *term contracts* for providing construction or other work when ordered by a promoter at any time within an agreed period (the *term*).

Under all such contracts the promoter in effect employs a contractor as an extension of his own organization. To control the cost of cost-reimbursable work the promoter's project team must direct the contractor's use of resources. The contractor's risks are limited, but so is his prospective profit.

The advantages of cost-reimbursable contracts are as follows.

- They provide extreme flexibility.
- They provide fair payment for and good control of risk.
- They allow and require a high level of promoter involvement.
- They facilitate joint planning.
- Knowledge of actual costs is of benefit to the promoter in estimating and control and in evaluating proposed changes.

The disadvantages of cost-reimbursable contracts are as follows.

- There is little incentive for the contractor to perform efficiently.
- There is no estimate of final price at tender.
- Administrative procedures may be unfamiliar to all parties. In particular, the promoter must provide cost accountants or cost engineers, who must understand the nature of a contractor's business.

The uses of cost-reimbursable contracts, which are weak contracts from the promoter's viewpoint, are restricted to contracts containing major unquantifiable risk and to projects where no other form of contract is feasible. They can be used when it is impossible to define the quantity of work and the programme: offshore hook-up contracts are good examples of this circumstance.

There are a number of situations when the use of this type of contract can be justified, when the work is of an emergency nature or when the work is innovative and productivities are unknown, for example involving research and development. Cost-reimbursable contracts can be used if the work is of exceptional organizational complexity: for example, when there are multi-contract interfaces, to the extent that definition of a target cost is impossible. A common use is on occasions where a contractor is required to rescue or complete a project that has been subject to extensive disruption.

Target cost

This is a development of the reimbursable type of contract whereby a promoter and a contractor agree at the start a probable (or target) cost for a then uncertain scope of work. The contractor's actual costs are monitored and reimbursed as in a cost-reimbursable contract. Any difference between the final actual cost and the final target cost is shared by the promoter and the contractor in a way that is defined by the incentive mechanism. Before reaching the stage of tender documents, the promoter is recommended to investigate the implications of several different incentive mechanisms. If the incentive is to be maintained the target cost will subsequently be adjusted for major changes in the work and cost inflation. A fee, which is paid separately, covers the contractor's overheads, any other costs specified in the contract documents as not being allowable under actual cost, and his profit.

The following simple example and Figure 12.3 show the effects of a 50/50 share incentive mechanism and a fixed fee on the payment by the

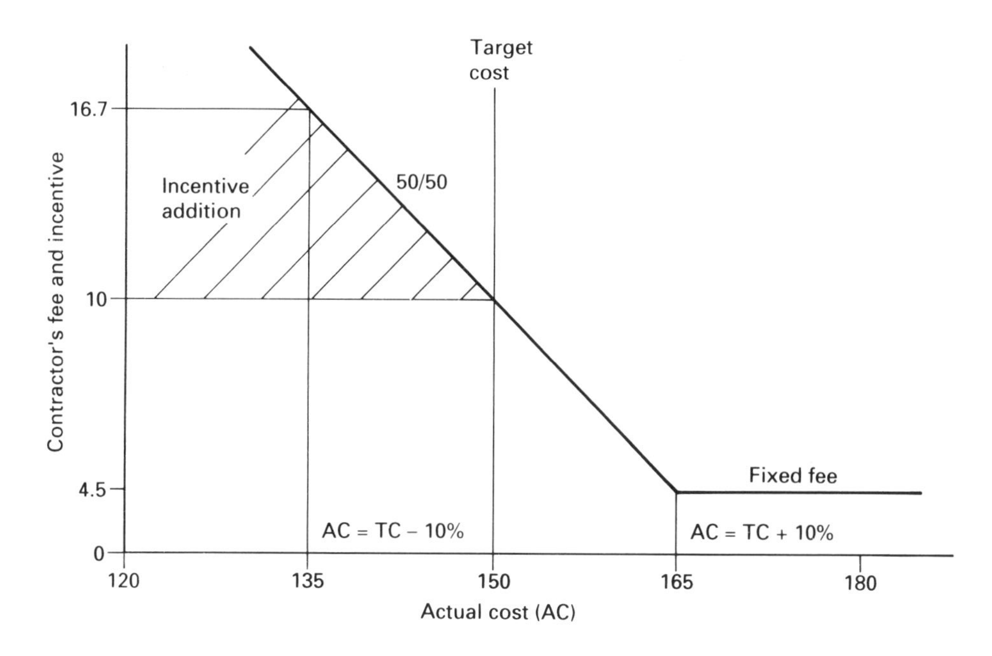

Figure 12.3 Target cost mechanism.

promoter and on the financial incentive to the contract. The target cost is £150 m, sharing of difference between actual cost and target cost is 50/50 promoter and contractor, and the fixed fee is £15 m.

In the case where actual cost equals £135 m, the total payment by the promoter is £157.5 m which is less than the sum of £165 m that he would have paid if actual cost had equalled target cost. The contractor receives a lower total payment, but his margin, as a percentage of actual cost, is increased from 10% to 16.7%. Conversely, if actual cost exceeds target cost the promoter is partly cushioned by the incentive mechanism. The cushioning effect also applies to the contractor, but as he has to bear 50% of the excess costs, his margin is much less attractive (4.5%). This illustration clearly demonstrates two powerful features of a target cost contract: first, the motivational effect of the incentive mechanism on the contractor, and second, the benefits to both parties of working together to keep actual cost under strict control and to bring it below target cost whenever possible. Thus the target cost mechanism remedies the principal weakness of a pure cost-reimbursable contract by imposing an incentive on the contractor to work efficiently.

The advantages of this form of contract are as follows.

- It provides fair payment for and good control of risk.
- It provides a high level of flexibility for design change.
- There is identity of interest: both parties have a common interest in minimizing actual cost. Fewer claims result and settlement is easier.
- There is promoter involvement: the contract offers an active management role for the promoter or his agent. Joint planning aids integration of design and construction, efficient use of resources and satisfactory achievement of objectives.
- It is realistic: the facility to require full supporting information with bids, coupled with thorough assessment, ensures that resources are adequate and the methods of construction are agreed.
- Knowledge of actual costs is of benefit for estimating and control, and in evaluating proposed changes.

There are some weaknesses in using this type of contract, which might make it unsuitable for certain projects; important points to consider include the following.

- Promoter involvement is essential and he must take a different attitude from that adopted on the price-based contracts.
- Unfamiliar administrative procedures and a probable small increase in administration costs. In addition to cost accountants some measurement engineers will be required for purposes of target adjustment. The initial target cost provides no greater certainty about final price than the tender total in an admeasurement contract.

The use of this form of contract should be considered when there is inadequate definition of the work at time of tender owing to emphasis on early completion or an expectation of substantial variation in work content, when the work is technically or organizationally complex or when the work involves major unquantifiable risks. The role of the promoter is significant, and a target contract can be used if the promoter wishes to be involved in the management of the project or wishes to use the contract for training of his own staff or for development of a local skilled construction labour force.

Targets can be based on other project parameters instead of, or as well as, cost. Time target contracts are popular for certain types of work and Figure 12.4 shows a graphical representation of a combined time and cost target mechanism.

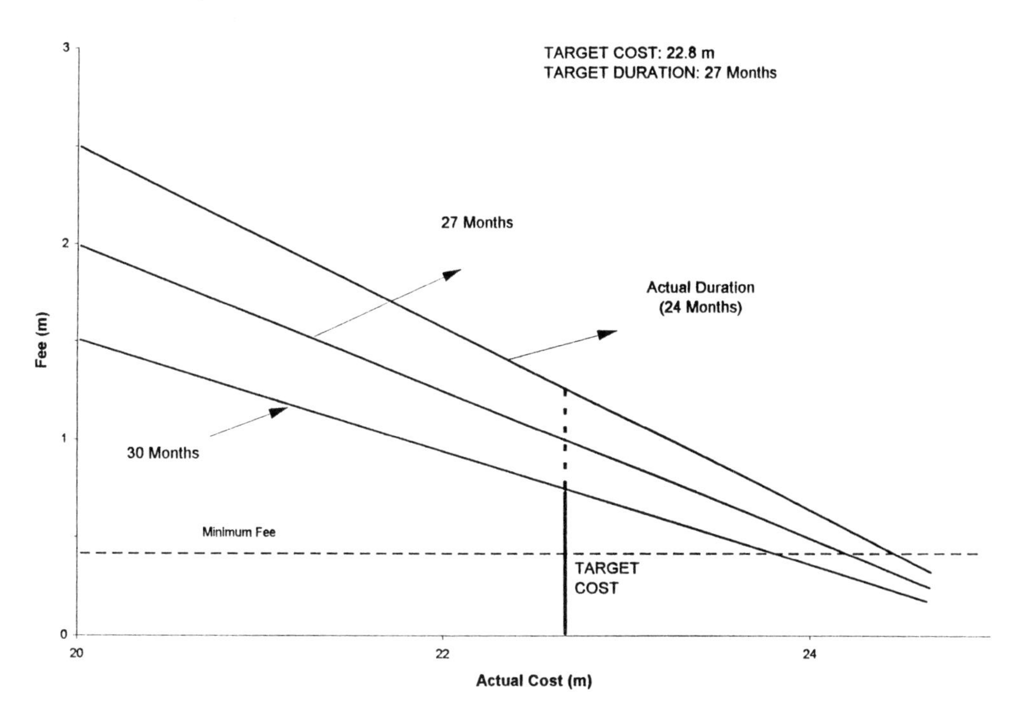

Figure 12.4 Combined time and cost target.

12.8 Model or standard conditions of contract

The significance of model forms of contract has been mentioned in section 12.2. There are a variety of model forms, and Table 12.1 shows some of the forms commonly used in the UK for engineering projects. These define the legal conditions under which the parties agree to contract.

Project managers can use private, in-house contracts or commission new contracts for a specific project. There are drawbacks with this approach: primarily that the contract will not have been 'tested at law' and the precise meaning and interpretation of any clause in a new contract could, potentially, be the source of lengthy litigation. It is more likely that the project manager would select the conditions of contract that most closely match the requirements and then modify clauses or add additional 'special' clauses as necessary.

The most recent and more fundamental innovation in the UK has been the initiation of the New Engineering Contract (NEC), published in 1995. This has been sponsored by the Institution of Civil Engineers, but is intended for all types of engineering and construction project. The

Table 12.1 British model forms of contract.

Institution of Chemical Engineers Conditions of Contract for Complete Process Plants for Lump Sum Contracts, revised edition 1981. Known as the IChemE 'Red Book'.

Conditions of Contract for Complete Process Plants for Reimbursable Contracts, Institution of Chemical Engineers, 1976 (revised edition 1992). Known as the IChemE 'Green Book'.

General Conditions of Contract, model form MF/1, the Institution of Mechanical Engineers, the Institution of Electrical Engineers and the Association of Consulting Engineers, 1995 (replacing previous model forms A and B3).

Conditions of Contract for Civil Engineering Works, Institution of Civil Engineers, the Association of Consulting Engineers and the Federation of Civil Engineering Contractors, 6th edition, 1991. Known as 'the ICE 6th edition'.

Conditions of Contract for Minor Works, 1988, Institution of Civil Engineers.

GC/Works/1 Revision 3 Conditions of Contract for Construction, prepared by the Property Services Agency and published by HM Stationery Office, 1988.

JCT Standard Form of Building Contract, Joint Contracts Tribunal for the Standard Form of Building Contract (JCT), 1980. Alternatives for private and local authority promoters, and with and without quantities or approximate quantities. Known as 'JCT 80'.

JCT Intermediate Form of Building Contract, Joint Contracts Tribunal for the Standard Building Contract (JCT), 1984 (for smaller building projects).

JCT Management Contract, Joint Contracts Tribunal for the Standard Building Contract (JCT), 1987.

Works Contract 3: Joint Contracts Tribunal for the Standard Building Contract, 1987 (for use with the JCT Management Contract).

The New Engineering Contract Document, The Engineering and Construction Contract 1995, Institution of Civil Engineers.

NEC consists of relatively brief core terms for all contracts, and modules of optional terms for use with the core terms as needed, for instance to make a cost-based rather than price-based contract. With this flexibility it could replace the variety of models at present in use in the UK and other countries. Its flexibility could enable promoters to cease using their own conditions of contract and end the UK and US practice of amending and adding to model forms.

One aim of the NEC is to reduce the extent of disputes that arise from current contracts. The NEC is designed to be used within a variety of legal systems including the English law system and be simpler to read and understand than most other models, which should help non-UK users.

12.9 Tendering procedure

Several different procedures are used for selecting vendors, suppliers, tenderers and contractors:

Competitive: open or select (a restricted number of bidders).

Two stage: a bidder is selected competitively early in the design process. The tender documents contain an outline specification or design and approximate quantities of the major value items. As design and planning proceeds the final tender is developed from cost and price data supplied with the initial tender.

Negotiated: usually with a single organization but may be up to three.

Continuity: tendering competitively on the basis that bidders are informed that the successful party may be awarded continuation contracts for similar projects based on the original tender.

Serial: the bidder undertakes to enter into a series of contracts, usually to a minimum total value. A form of standing offer.

Term: the bidder undertakes a known type of work, but without knowing the amount of the work, for a fixed period of time.

In all cases a prequalification procedure may be adopted.

12.10 Factors affecting strategy

In making the decision about the choice of contract strategy a number of factors should be considered. Clear definitions of the promoter's objectives are required so that the significance of these factors can be established.

The responsibilities of the parties need to be determined. The responsibility for defining design, quality, operating decisions, safety studies, approvals, scheduling, procurement, construction, equipment installation, inspection, testing, commissioning and for managing each of these must be established. Related to this the risks must be allocated between the parties: who is to bear the risks of defining the project, specifying performance, design risks, selecting subcontractors, site productivity, mistakes and insurance? The terms of payment for design, equipment, construction and services are a major influence on the choice of contract strategy.

It is also important to consider the provision of an adequate incentive for efficient performance from the contractor. This must be reflected by an incentive for the promoter to provide appropriate information and support in a timely manner. The contract may need to be flexible. The prime aim is to provide the promoter with sufficient flexibility to introduce change that can be anticipated but not defined at the tender stage. An important requirement is that the contract should provide for systematic and equitable evaluation of such changes.

The interrelationship of these requirements with the type of contract is demonstrated in Figure 12.1. The requirements are expressed in terms of contractor's incentive, promoter's flexibility and contractor's risk contingency. It is apparent that, generally, contractor's incentive and promoter's flexibility tend to be incompatible. For example, a lump sum contract imposes maximum incentive on the contractor but also implies a very high level of constraint on the promoter against introducing change. The converse is true at the other extreme of a cost-reimbursable plus percentage fee contract. However, the target cost contract offers greater flexibility than the admeasurement contract while providing a similar level of incentive.

Many or almost all of these factors could be important on a project, but it is likely that certain factors will dominate. All the primary factors are significant but they also conflict. If working on an offshore energy project 'time' might be the dominant objective to meet a weather window or to avoid disruption of the flow of crude petroleum; this would have obvious implications for the cost of the project.

Primary:	minimum cost, timely completion, quality, risk
Secondary:	promoter involvement, co-operation, capital, actual cost
Tertiary:	technology, market, industrial relations, disciplines
Overseas:	design, appropriate, plant sales, risk

There are a number of possible strategies that could satisfy the identified objectives, and it is the task of the project manager to advise the promoter, in the knowledge of the clearly defined project objectives, which strategy to adopt. This selection is possibly one of the most important decisions in any project.

There can be other criteria to consider, and every project has to be assessed individually. For example, changes and innovations in contract arrangements have followed the privatization of what were formerly public services in the UK, to try to meet commercial rather than political accountability.

12.11 Subcontracts

The same contract strategy principles apply to decisions made by a main contractor to employ subcontractors: hence the section above can be interpreted at a number of levels with different parties filling the 'promoter' and 'contractor' roles. For example, in a subcontract the contractor fills the promoter role and the subcontractor fills the contractor role, but in a further supply contract the subcontractor could fill the promoter role and the vendor fills the contractor role.

A common principle is that in a main contract a contractor is responsible to the promoter for the performance of his subcontractors. Practice varies in whether a main contractor is free to decide the terms of subcontracts, choose the subcontractors, accept their work and decide when to pay them. It also varies in whether and when a promoter may bypass a main contractor and take over a subcontract.

In nearly all engineering and construction the main contractors employ subcontractors and suppliers of equipment, materials and services in parallel.

Further reading

Allwright, D. and Oliver, R.W. (1993) *Buying for Goods and Services*, Chartered Institute of Purchasing and Supply, London.

Barnes, N.M.L. (1991) The New Engineering Contract. *International Construction Law Review*, **8** (2), 247–255.

Merna, A. and Smith, N.J. (1990) Project managers and the use of turnkey contracts. *International Journal of Project Management*, **8** (3), 183–189.

NEDO (1982) *Guidelines for the Management of Major Projects in the Process Industries*, National Economic Development Office, London.

NEDO (1982) *Target Cost Contracts – A Worthwhile Alternative*, National Economic Development Office, London.

Perry, J.G. and Thompson, P.A. (1992) *Engineering Construction Risks – A Guide to Project Risk Analysis and Risk Management*, Thomas Telford, London.

Wright, D. (1994) A 'fair' set of model conditions of contract – tautology or impossibility? *International Construction Law Review*, **11** (4), 549–555.

Wright, D. (1994) *Guide to IChemE's Model Forms of Conditions of Contract*, Institution of Chemical Engineers, Rugby.

Chapter 13
Tender Procedures and Contract Policy

Any organization that wishes to employ others to undertake work on its behalf must have a coherent policy for their selection. This chapter examines and proposes strategies for the identification and selection of the most appropriate policies for the prequalification of contractors or suppliers and for the evaluation of competitive tenders. Following the project stages in chronological sequence, the text goes on to describe methods of measuring the work to be undertaken and mechanisms for pricing variations.

13.1 Contract policy

It is important that every promoter involved in contracting has a formalized contract policy established by management, broadly defining why, what, when and how work should be contracted out by the promoter.

The objective of going out to contract is to obtain specific works, services and goods required to support the promoter's general business objectives that cannot or will not be supplied by the promoter for various reasons, such as: the contractor can supply the works, services or goods at lower cost/risk than the promoter; the contractor can supply the works, services or goods with staff, labour and expertise that cannot or cannot sufficiently be made available from the promoter's own resources; and the type, quality, enhancements and fluctuation of required works, services or goods are inherent to the contractor's specialism and not to the promoter's. Contracting should encourage the optimum utilization of the promoter's own executive staff and keep administrative overheads to a practical minimum.

As a general rule contractors should be independent, self-sufficient and 'at arm's length', and the promoter's aim should be to manage the contract, not the contractor.

Matters that merit review for a consistent policy approach include:

- qualification of contractors by qualities such as reputation, experience and reliability;
- circumstances under which negotiation, rather than competitive tendering, may be appropriate and/or acceptable;
- types of contract to be preferred in given circumstances;
- facilities and services that may be provided to contractors by the promoter;
- commercial aspects included in the contracts themselves.

Procedures

Contract procedures are written for the purpose of outlining the objectives and scope of contract policy and the manner in which it is to be applied. They are designed to ensure that:

- a promoter's policies and procedures are clearly stated and applied;
- the authorities and responsibilities for the preparation, award and control of contracts are defined;
- appropriate input from advisory functions is incorporated at each stage of the contracting process;
- information passes in an orderly manner to those parts of the organization, as well as to and from appointed contractors, on a need-to-know basis;
- the best interests of the promoter and its staff are protected and safeguarded.

Business ethics

Management should establish a code of conduct, which is brought regularly to the attention of all staff, covering:

- the general policy to be adopted in relationships with contractors and their staff, who should be dealt with in an equitable and business-like manner;
- the obligation to declare any conflicts of interest, potential as well as actual, should they arise;
- the importance of confidentiality and security in all matters concerned with contracting activities.

Contractors should also be made aware of promoter policies with regard to conflicts of interest and the giving and receiving of gifts etc., and promoters may consider it appropriate to incorporate a suitable clause on the subject in their general conditions of contract.

13.2 Contract planning

During the initial phases of a project, discussions should take place under the direction of the project manager between the various disciplines concerned to formulate, review and develop the different aspects of the project, culminating in the preparation of a project contracting plan. It will be necessary to prepare a number of plans to cover the detail of each contract to be let.

An overall schedule should be kept to monitor the critical dates in the preparation and letting of contracts over the whole project. Similarly a schedule should be kept to monitor critical dates within the sequence of activities for letting each contract. As these discussions proceed, attention will be given to the various implementation considerations including those factors that will have some influence on the type and number of contracts, selection of possible contractors, and method of operation.

The contracting strategy addresses the choice of the number and types of contract that will most effectively contribute to the success of the project with the least commitment of the promoter's resources. The types of contract chosen also affect the master plan because of the varying requirement for control and monitoring. Competitive tendering would be the normal method adopted for letting contracts, and the contract should be awarded to the lowest conditioned tender provided other criteria of comparability and acceptability are met.

Number of contracts

The first step in contracting planning is determining the number of contracts into which a project will be divided. One of the basic considerations must be the effect of the number of contracts on the promoter's management effort. More contracts mean more interfaces and greater management involvement. Fewer contracts reduce this involvement but may increase the promoter's risk exposure.

The following principles should be used in determining the number of contracts.

- Each contract must be of manageable size and consist of elements that the contractor can control.

- The contract size must be within the capacity of sufficient contractors to allow competitive tendering. Occasionally, only one contractor may be capable of undertaking certain types of work and competitive tendering would not be possible. Such commitment should be minimized.
- The time constraints of the work may require parallel activities, and this may mean that capacity restrictions necessitate separate contracts rather than a single contract.

In setting the number of contracts it is useful to list the elements of work and the contract phases from conceptual design to test and commission. These can be considered in a number of combinations to determine the minimum number of manageable contracts.

Tender stages

There are two stages at which the promoter and his project manager can control the selection of contractors: first, before the issue of tender documents, and second, during tender analysis, before contract award. Both evaluations are important but have different objectives:

Pre-tender: to ensure that all contractors who bid are reputable, acceptable to the owner and capable of undertaking the type of work and value of contract.

Pre-contract: to ensure that the contractor has fully understood the contract, that his bid is realistic and his proposed resources are adequate (particularly in terms of construction plant and key personnel).

13.3 Contractor prequalification

The principal choice is whether to adopt a full prequalification procedure specific to each contract or whether to develop standing lists of suitably qualified contractors for various sizes of contract and types of work. Any form of prequalification is not a straightforward matter as it involves subjective judgement.

A full prequalification procedure may include:

- press announcement requiring response from interested firms or direct approach to known acceptable firms;
- issue by the promoter of brief contract descriptions including value, duration and special requirements;

- provision of information by the contractor including affirmation of willingness to tender, details of similar work undertaken, financial data on number and value of current contracts, turnover, financial security, banking institutions, and the management structure to be provided, with names and experience of key personnel;
- discussions with contractor's key personnel;
- discussions with other promoters who have experience of the contractor.

The evaluation may be done qualitatively, for example, by a short written assessment by a member of the project manager's staff to narrow down the number of suitable contractors.

After the potential bidders have been interviewed and evaluated, the project manager should recommend a bid list for the promoter's approval. This should be a formal document, which provides a full audit trail for the selection process. Specifically it should discuss the reasons for inclusion and exclusion of each contractor considered and confirm that all of those selected for the bid list are technically and financially capable of completing the works satisfactorily.

It is the usual practice to prequalify about 1.5 times the number of contractors to be included at tender. This is often achieved by a combination of quantitative and qualitative methods, as no standard procedure exists.

The EU Works Directive

Companies are subject to the UK's obligations, under the Treaty of Rome, not to discriminate on the grounds of nationality against contractors, or suppliers of goods and/or services from elsewhere in the Union. For supplies and works contracts where the estimated value exceeds certain thresholds, the Treaty obligations are reinforced by Procurement Directives. Subject to certain limited exceptions the Directives include provisions on:

- publication of pre-information on procurement intentions in the *Official Journal of the EU*;
- publication for calls for competition in the *Official Journal*;
- prescribed award procedures;
- criteria for rejection of unsuitable candidates;
- permitted proofs of economic, financial and technical standing;
- non-discriminatory selection of tenderers;
- prescribed contract award criteria;

- publication of award details;
- debriefing of unsuccessful tenderers.

13.4 Contract documents

For a small amount of work it is usually sufficient for a contract to consist of a drawing and an exchange of letters. The drawing can show the location and amount of work. The materials can be specified on the drawing and the completion date, price and other terms that matter to the parties stated in a letter. If the agreement is the result of a series of letters and replies, what is known as *offer* and *counter-offers*, one final letter should state all that has been agreed and replace all previous communications so as to leave no doubt as to what has been offered and accepted.

To avoid these doubts on all but small projects the practice has evolved of stating the various terms of a contract in a set of documents. These can be lengthy, but some can be the same for many contracts and so do not have to be prepared anew each time. The set of documents traditionally used in UK engineering contracts is:

- drawings
- specification
- conditions of contract
- agreement

and perhaps a separate schedule or BoQ, which defines the scope of the work and forms the basis of the terms of payment.

The basis of a successful construction contract is established by the tender documents and any subsequent negotiation prior to the award of the contract. The contract is defined by the contract documents, which for a conventional engineering admeasurement contract may include:

- form of agreement
- special conditions
- general conditions
- specifications
- drawings
- priced bills of quantities or schedule of rates
- pre-contract minutes or correspondence.

The contract documents should be concise, unambiguous and give a clear picture of the division of responsibilities and legal obligations between the parties. Risks should be identified and clearly allocated.

Five main aspects of contract implementation are covered by the contract documents: legal, technical, financial, organizational and procedural.

The main roles of the documents are, at tender, to provide a common basis on which contractors can bid and against which their bids are assessed. Post-tender, the documents enable fair payment for work done; should facilitate the evaluation of change; set the standard for quality control; and provide redress for either party in the event of breach of the contract or a term of the contract.

Points to note include the following.

- the provision of liquidated damages does not ensure the control of time.
- the contractor's programme is not usually a contract document.
- the contractor controls cost and time by controlling resources; in the majority of construction contracts the contract documents provide little or no information in this area.

13.5 Bid review

For larger-value projects a dedicated bid review team will be required. This usually comprises a core of, say, three people supported by specialists reviewing particular aspects of the bid. A typical core team might include the project manager and the lead design consultant. Specialist support may be needed from other design consultants, a contractual/legal expert or a QA specialist.

The chosen contract strategy and form of contract will influence the review mechanism:

- *Lump sum.* With this type of contract the objective of the review is to find the bidder who offers the lowest price, best programme and yet meets the specification in terms of scope, quality, operability and economic maintenance. If the bidder is to have design responsibility, design capability will also be reviewed.
- *Reimbursable cost.* The objective with this type of contract is to find the bidder who offers the best commercial terms coupled with a project execution capability that gives the team confidence that it will meet the requirements.
- *Other contracts.* These will fall somewhere between these two

extremes in terms of the promoter's risks, and the bid team must make a judgement on which bid will provide the best value for money.

13.6 Bid evaluation

The project manager will open the bids from the various contractors on a given date and time. A systematic evaluation of the bids would include examination of the following:

- compliance with the contractual terms and conditions (tenders may be qualified by contractors);
- correction of bid prices (if errors detected in multiplying rates by quantity);
- screening of bids for detailed analysis;
- pre-award meeting (optional);
- selection of the best bid and recommendation to the owner for contract award.

Bid conditioning is the term given to the process of reviewing all of the bids so as to be able to compare like with like. Some organizations take a rigid stance and only compare tenders that conform fully with the inquiry requirements. Others consider that bidders may offer alternatives of genuine benefit to both parties and hence consider all submissions. In that event it is usual to require bidders to submit a 'conforming' bid as well as variants to allow comparisons to be made. Where direct comparison cannot be made, exceptions must be carefully noted.

A misconceived bid is a bid based on an error or misunderstanding on behalf of the contractor and should be easily identified in the bid evaluation. Suicidally low bids are automatically rejected by some promoters as they are usually difficult to administer. However, in some circumstances a very low bid will be a strategy for 'buying' the job for the contractor's own purposes, and commercial reasons. In the evaluation an attempt must be made to discover the contractor's philosophy. If the promoter is to accept the low price they will want to be sure that the project will be realized for this price. The pre-award meeting is the ideal time to discover the motives of the contractor.

Contract award recommendations

It is essential that the bid review team put together a formal recommendation for award of contract. The report should:

- explain the background;
- summarize the recommendations;
- describe bid opening and the initial position;
- describe the bid-conditioning process;
- give reasons for rejection;
- summarize reasons for recommendation;
- indicate cost and time implications for the project.

13.7 Typical promoter procedure

The tender packages are often prepared by an organization employed by the promoter or by the consultant engineer responsible for the design. In either case the procedures to be followed are similar, although the titles of the departments and individuals concerned may differ between organizations. In large projects such as those implemented offshore the level of project management required often makes the services of a management contractor cost-effective. For the sake of conformity the procedures will be described as if a management contractor has been employed by the promoter to coordinate the project.

The project manager controls the tender package preparation possibly as head of a department with duties solely relating to the preparation of contracts.

The main purpose of the contracts department is to check the following:

- that no gaps or overlaps exist between the individual subject contracts;
- that all the particular requirements of the promoter are included and covered in the subject contracts.

In particular, the contracts department liaises and coordinates with the other departments with regard to matters such as:

- engineering
- QA/QC
- materials/procurement
- construction
- legal and insurance
- planning and scheduling
- costs and estimates
- accounts and payment.

The individual tender packages are usually developed from the pro-forma documents by selection of alternatives and/or amendments. These departments in turn should liaise with their counterparts in the promoter organization.

Draft formats of the individual subject contracts should be circulated to each department for review, allowing sufficient time to incorporate amendments within the tender package. A standard procedure should be adopted for the reporting of comments and amendments to the contracts department.

The final draft of the tender package is then sent to the promoter for formal approval prior to issue of tenders.

The contracts department maintains a register of tender documents, including information such as:

- reference numbers
- description of requirements
- date of issue to the tendering contractors
- date of return of tenders and the date that tenders remain open to.

The procedures and pro-forma documents used by the contracts department in preparing tender packages should be set out in a document (sometimes referred to as a *works plan*).

Invitation to tender

This is in the form of a letter to the tendering contractor written on behalf of the promoter.

The letter simply invites the contractor to submit a tender for the performance of certain work. The letter lists the tender documents attached and requests the contractor to acknowledge receipt of the documents and their willingness to submit a tender.

A *form of acknowledgement* is usually included with the letter so that replies from all tenderers are set out in a standard way.

Instructions to tendering contractors

The instructions to tendering contractors inform the tenderers what is specifically required of them in their tender and usually comprise the following items:

- tendering procedures
- commercial requirements

- information to be submitted with tender.

It must be made clear to tendering contractors that the tender submission should be based on the 'scope of works' contained in the contract documents. Alternative tenders are submitted with a conforming tender as set out in the original tender documents so that the promoter may compare the two. Where the promoter will not consider alternative tenders, this should be expressly stated in the instructions to tendering contractors.

Conditions of contract (articles of agreement)

The conditions of contract can be prepared by the contracts department in consultation with the legal and insurance disciplines. They are prepared as a standard for the whole project and must be agreed with the promoter.

The conditions of contract are often based on an industrial standard or the promoter's standard. In either case it is likely that modifications will have to be made to suit the unique requirements of the project, the structure of the contract package and the philosophies of the contracting procedures.

A *form of agreement* for executing the contract and identifying the contract documents is attached to the conditions of contract.

The standard conditions can be modified as necessary for the individual contract packages.

The promoter and the contractor are the parties to the contract and as such are legally bound to abide by the conditions set out. A project manager may be appointed by the promoter, via a contract of employment, to administer the contract and to monitor the performance of the works, and does not have any contractual link with the contractor.

Brief description of the works

The brief description of the works explains the overall requirements and parameters of the works to be performed. The description should be drafted with the aim of creating a broad appreciation of the works to be performed. The overall dimensions of the project should be stated and technical links between the elements of the contract package should be described.

The description should be kept brief and simple and not repeat detail already covered by the drawings, specification, programme or any other work element of the contract package. It should define any work that is not covered elsewhere in the contract package. The interface with other contract packages should be described without reference to the actual work contained.

This section may also be used to detail such things as the facilities that are required by the project's site team.

Programme for the works

For larger projects the programme is usually prepared by the planning and scheduling department, liaising with the contracts department. It should contain all key dates for the particular contract, including:

- award of contract;
- start of the works;
- completion of the works;
- intermediate key dates for particular elements of the work, where such elements are required to interface with work outside the scope of the subject contract;
- critical dates within the contract.

Indexes of drawings and specifications

These are usually prepared by the engineering department. The indexes should be correctly numbered to match the latest numbers and revisions of drawings and specifications contained in the tender packages.

The format, but not the contents, of the indexes should be prepared as a standard by the contracts department in consultation with the engineering department.

Drawings and specifications

These are prepared by the engineering or quality control and assurance departments. Generally, two copies of each drawing and specification are included in the tender package for each bidding contractor.

The terminology of the specifications must be consistent with that of the contract documents. The contracts department should review all specifications before issue to ensure that the terminology is consistent. Defined terms are always started with capital letters. The contracts department should check the individual parcels of drawings and specifications against the indexes in the tender documents. The specifications and drawings issued with tenders (i.e. those listed in the index and specifications and drawings) should be given an identifying revision number.

The copying/printing and binding of the drawings and specifications is then coordinated by the contracts department.

Promoter-provided items

Where items are to be supplied by the promoter they must be listed. No specific mention should be made of those items to be provided by the contractor. Reliance is placed on the phrase 'all other items to be provided by the contractor'.

Practical delivery periods should be stated against each item. Note that such periods become contractual commitments and should be strictly adhered to.

The format of the promoter-provided items should be developed by the engineering department in liaison with the contracts department. The following items are usually considered:

- descriptions of the items provided
- delivery periods
- specific storage requirements
- explanation of any markings
- details of returnable packaging.

Contract coordination procedures

The contract coordination procedures describe the administrative requirements for the implementation of the contract. The procedures explain the day-to-day duties and responsibilities of the site management team and the contractor's site team. The procedures also detail the lines of communication. Guideline procedures detailing the implementation of the technical requirements can also be included, but to avoid inconsistencies a check must be made for any duplications or contradictions with the specifications.

In order to compile the coordination procedures the construction department works with the various disciplines to obtain their requirements for the implementation of their specific responsibilities.

Form of tender

This is prepared by the contracts department and provides a format whereby the contractor confirms:

(1) examination of all the tender documents including notices given during the tender period;

and whereby the contractor offers:

(2) to perform the works;
(3) to keep the tender open for a defined period;
(4) to enter into an agreement;

and whereby the contractor understands:

(5) that the tender is not an authority to proceed;
(6) that the promoter or management contractor are not liable for any costs;
(7) that neither the lowest tender nor any offer need be accepted;
(8) that only part of the works may be awarded;

and provides for the contractor's authorized signature.

An appendix is usually attached detailing the following:

(1) amount of performance bond when required;
(2) minimum third party insurance;
(3) date of commencement of the works;
(4) date of completion of the works;
(5) liquidated damages;
(6) period of maintenance;
(7) limit of retention monies;
(8) time for payment after receipt of invoice.

Contract price

This is prepared by the contracts department with the assistance of the cost control department.

The pricing format allows for all anticipated items of work to be performed by the contract, such as:

(1) general obligations of the conditions of contract;
(2) contractor's facilities;
(3) working in accordance with the coordination procedures;
(4) promoter's/engineer's office accommodation;
(5) promoter's/engineer's transportation;
(6) stores, material handling and waste;
(7) temporary works;
(8) provisions of special items;
(9) observance of any restrictions;

(10) all disciplines – structural, outfitting, cladding, pipework, equipment, surface preparation and protective coating, thermal insulation, fireproofing, weighing, electrical, instrumentation, load out and seafastening.;
(11) dayworks and contingencies;
(12) provisional sums;
(13) contractor-provided items;
(14) transportation of the works;
(15) escalation;
(16) currency fluctuation.

The pricing format incorporates a fair method of assessing interim payments and the valuation of variations, and also provides a method by which the costs can be controlled for budget forecasting and cost allocation.

Use of pro-forma documents

The pro-forma document used as a draft for the preparation of the individual subject tenders and contracts usually comprises the sections listed in Table 13.1.

Assembly of tender package

On completion of all the information the final Tender package is assembled by the contracts department. As a general rule the package consists of the invitation to tender, form of acknowledgment and the documents listed in Table 13.2. Standard covers are usually prepared by the contracts department.

Review of the tender package

Following the assembly of the tender package a copy is sent to the department managers of the management contractor for review and comments. These reviews and comments must be completed within the time-scales laid down in the contracting work plan, and time limits should be stated on the transmittal.

Following the reviews any comments should be incorporated where necessary. Reasons should be given by the contracts department for not incorporating any comment. At least two copies of the reviewed tender package should be sent to the promoter for review and comments. As

Table 13.1 Use of pro-forma documents.

Section	Pro-forma used as standard	Individual sections to be prepared
Invitation to tender	√	
Instructions to tendering contractors	√	
Conditions of contract	√	
Brief description of the works		√
Programme for the works		√
Index of specifications	(format)	√
Index of drawings	(format)	√
Owner-provided items	(format)	√
Contract coordination procedures	√	
Form of tender	√	
Contract price		√
Form of agreement	√	

above, these must be completed within the time-scales laid down in the contracting work plan.

Collation and issue of tender packages

Following the final approval of the tender documents from the promoter the tender packages should be collated by the contracts department. 'Sealed package labels' should be provided for the return of the tenders from the tendering contractors.

The final collation of the tender packages and the issue to the individual tendering contractors should be in accordance with the contracting schedule, and is normally the responsibility of the contracts department.

Queries from tendering contractors

All queries from the tendering contractors during the tender period should be answered by the contracts manager. Any other contact

Table 13.2 Tender package.

Volume	Title	Prepared
I	Instructions to tendering contractors	Contracts
	Conditions of contract	Contracts
	Brief description of the works	Engineering/Contracts
	Programme for the works	Control/Contracts
	Index of specifications	Engineering/Contracts
	Index of drawings	Engineering/Contracts
	Promoter-provided items	Engineering/Contracts
	Contract coordination	Construction/Contracts
II	Form of tender	Contracts
	Contract price	Contracts/Control
III	Specifications (as indexed in Volume I)	Engineering/Contracts
IV	Drawings (as indexed in Volume I)	Engineering

between the tendering contractors and the promoter or other members of the staff during this period is to be discouraged.

13.8 Post-contract award

The contract will be awarded on the basis of conditions of contract containing an agreed payment mechanism: this covers the works as detailed in the drawings and specifications and any variation to it. The methods to be used for valuing the work are usually stated in the contract documents; there may be a number of different methods within the contract for different types of work, depending on the pricing formats used and the complexity of the work.

Payment mechanisms that may be used are lump sum, unit rate or cost reimbursable. Note that a lump sum contract does not necessarily have variations priced on a lump sum basis; the unit rate and/or cost reimbursable may be used if more appropriate. Similarly, variations under an admeasurement contract need not be priced using unit rates. Valuation of variations or of the original works specification relates to the method used, and the mechanism is the same whether it is a variation or permanent works.

Lump sum

The term 'lump sum' refers to the total price calculated for performing a specific piece of work; it does not have to be related to the way that the sum is calculated. The calculation of the lump sum may be based on:

- number of man-hours × rate
- number of plant-hours × rate
- cost of materials + mark-up
- subcontractor's cost + mark-up
- measured items of work × unit rate.

It is important that the scope of work is clearly defined, because the details of the price may not be available for examination by the promoter: for example, the main contractor may not include for the work of a nominated subcontractor. Once a lump sum price is agreed it has to be assumed that the price will not be adjusted for any reason.

Unit rate

The unit rate is the cost for performing a specific item of work, and is commonly found in the BoQ. All rates are related to the quantities of permanent work. The work to be included in each rate will depend on the rules defined in the relevant documents; these rules also cover the method of measuring the item. In civil engineering the Civil Engineering Standard Method of Measurement 3rd Edition 1991 (CESMM3) applies. In the oil industry there is no standard.

The cost of supplying materials may be included in the unit rates but alternatively the cost may only be for handling material supplied by the Promoter.

The rules describing the method of measurement and the instructions about what is to be included in the rate must be examined in detail for every job.

Reimbursable

This is usually adopted when the other methods are not considered to be appropriate; it is typically used for complex or risky work. Contracts usually include rates for man-hours for various categories of personnel and rates per hour for various types of plant. The actual hours used are recorded and applied to the rates. Costs of materials are usually reimbursed at cost plus a mark-up. This method leads to uncertainty as to final cost, and there can be a lack of incentive for the contractor to perform the work efficiently.

The contractor prepares forms specifying the actual work, the hours spent, the type of labour, the materials incorporated in the work and the plant used. The forms are signed by an authorized signatory for both parties to record agreement on the resources spent on the work.

13.9 Valuation procedure for variations

Generally, under an admeasurement or unit rate contract, if the varied work is of a similar nature to existing items of work it is valued at existing rates and prices. If there are no similar rates or prices then the existing ones are used as a basis for valuation and new rates are agreed by the promoter and contractor. If no agreement can be reached then new rates are fixed; this can be disputed, in the form of a claim, by the contractor.

Some conditions of contract require the contractor to submit a quotation in the event of a contractual change for the varied work. The basis for pricing this could relate to any of the categories discussed in the previous section.

Under a lump sum contract the pricing of a variation will be very difficult unless a schedule of rates has been submitted by the contractor, whereas under a reimbursable contract any variations are simply measured and valued in the same way as the rest of the work: that is, at cost.

Further reading

Ashworth, A. (1986) *Contractual Procedures in the Construction Industry*, Longman, London.
Barnes, N.M.L. (1992) *The CESMM3 Handbook*, Thomas Telford, London.
Boyce, T. (1992) *Successful Contract Administration*, Thomas Telford, London.

Part III

International Projects

Chapter 14

Multidisciplinary, Industrial and International Projects

The current demand for higher quality of living and greater technological development has meant that large engineering projects are increasing in number and complexity year by year. The unique nature of each individual project makes successful project management a difficult task. These projects can be regarded as investments with significant risk, organizational and engineering implications. This chapter concentrates on this type of project management and offers general guidance for the project manager for both domestic and international projects.

14.1 Multidisciplinary projects

The current world market recession has placed increasing demands upon the need to manage and deliver multidisciplinary projects in a more effective way. This has inspired a plethora of project management approaches: some new, and some revised versions of existing ideas. The more complex the projects and the greater the number of parties involved, the greater the corresponding need for better project management, although little progress has been made.

One problem in particular is causing increasing difficulty. It concerns the abilities of large organizations, which act in a number of roles – promoter, contractor, operator, processor, supplier and financier – and tend to be involved in a large number of concurrent projects which are themselves composed of many sub-elements, to manage effectively. This combination of many complex projects occurring simultaneously causes new problems for the project manager. The oil companies are prime examples of this type of organization.

This is compounded by the fact that it is difficult to identify the significant problems and risks inherent in multi-project engineering management and to balance the cost-effectiveness of corporate management methods against new information technology developments.

It is always difficult to prejudge the evolution and development of existing systems, particularly in the more subjective management areas, but the increasing size of multinational companies means that multi-project management is likely to become more significant. The increasing size of multinational companies also means that this subject becomes more important, and if the trend to replace strategic company management with the project management of a large number of concurrent internal and external projects continues, then progress in this area becomes essential.

Currently, multi-project management is still regarded as an unsatisfactory balance between sophisticated project management software packages, which can view a large number of concurrent projects and optimize resources, investments and returns but without being able to incorporate the strategic company or organizational needs, and conventional corporate decision making at the highest level, which is providing long-term objective and subjective inputs to determine goals but cannot assess the implications, complexities and contradictory demands of large numbers of projects. Project management has to address the integration of these two separate levels in the traditional management structure. Flatter, more flexible organizational structures, as discussed in Chapter 9, are more suited to a project environment, but determining the priorities of tactical and strategic objectives in multi-project management remains the key issue.

The existing approaches seem to be either technically based or management based, and there is mutual distrust and misunderstanding of the relative strengths and weaknesses of the two approaches. These companies invest large amounts of money each year in new projects, and it is likely that even small savings achieved by improving multi-project management would produce large returns which could be reinvested in new projects.

For multi-project management the management techniques are the subject of research in many countries. There is a need to address the areas of processing of information, decision analysis, intergroup communication, command structure and interpersonal relationships. The multi-project environment is unique and affects all the areas of the management system and its structure. New information is required about the structure of the multi-project environment and the effectiveness of the 'triage' concept of multi-project management decision making indicating the need to change the management process and priorities with time.

14.2 Industrial projects

The number of industrial engineering projects undertaken each year is steadily increasing and seems likely to continue to do so for the foreseeable future. Many reasons for this trend have been proposed but there is some agreement that the most significant factor is the relatively recent combination of an increasing world population with the consequent demands for energy and raw materials, the enhanced capabilities of engineers and the availability of private sector investment finance.

For most projects a comprehensive contract is usual, based on limited competition and then negotiation of a specification, scope of work, price and other terms between promoter and contractor, sometimes including rights to licensed processes. International promoters usually specify their own detailed design requirements and standards, but otherwise the contractor is responsible for design and construction to achieve performance, including subcontractors' and vendors' items unless otherwise negotiated, but subject to supervision. Design is dependent upon detailed information from systems subcontractors and vendors.

The orders to vendors in other countries for equipment mostly follow the promoters' own conditions of contract, but some use the model conditions published by the UN Economic Commission for Europe.

Usual practice is to adopt contracts with fixed price stage payments to the main contractor, less retention. The contractor might be entitled to extra time and extra payment for a few given conditions. Many contracts combine payment on the fixed price basis for definite work and a day-works schedule of rates per hour or day for paying for the use of manpower, machines, etc. when ordered.

Occasionally, entirely cost-based contracts are used, usually when the scope or the nature of the construction work is undefined when entering into the contract. For these one comprehensive contract is usual, based on a specification of the resources that the contractor will have to provide and classes of work that he may have to undertake, usually written by the promoter in consultation with the prospective contractors. Instructions from the promoter then define the scope, quality and timing of the work to be done, usually progressively as decisions are made after construction has started.

Sometimes the promoter's and the contractor's project management and design teams are combined. The basis of payment to the contractor is not standardized; typically a combination of fixed price and reimbursable elements monthly payment is normal, less retention.

14.3 Large engineering projects

It is necessary to identify the difference between a large engineering project and a complex and expensive conventional engineering project. The terms 'large', 'jumbo', 'super' and 'giant' are all used in project management literature, and whilst not a problem of semantics the objective is to make clear that these projects have specialist management problems in addition to those encountered in conventional engineering management. Giant projects are so difficult to manage successfully that they should be restricted to those projects that cannot be realized in any other form.

An alternative approach to identifying large engineering projects is based on the concept of projects as long-term financial investments with considerable engineering implications. These investments should be optimized, which may result in the design and construction phases of the project being executed at above minimum cost in order to satisfy project objectives.

The management of large engineering projects has not been particularly successful; many projects have been abandoned at various stages of completion. Many of the projects that are completed break even and only a few are profitable; moreover, the major losses and gains do not usually occur in the operation phase of a project but in the project start-up and commissioning. Therefore the methods and techniques applicable to giant project management would appear to be significantly different.

For large projects, finance is important, and there should be strong credit support from the project promoters. Although this will usually form a minor share of the total financing package, 20–30%, it demonstrates the confidence of the promoters in the project. Investors would be unwilling to consider a project that had little equity capital; project risks, technological, supply of materials, debt service, *force majeure* are usually regarded as acceptable but equity risks, creditworthiness, indirect completion guarantees, security, are not.

There are immense differences between public and private large engineering projects. In the public sector projects are selected, by agreement, from a priority list drawn up within a corporate national development framework. The use of appropriate technology is reviewed at the design stage, and appraisal for loan approval is made on the basis of a combined technical, institutional, economic and financial assessment. Only then are competitive bids sought from contractors. In contrast the private sector selects projects on the criteria of individual judgement and market principles. A more thorough evaluation of the financial suitability is required as the project is not state guaranteed, and a higher proportion of the funding may be supplied by the project promoter with the risk of bankruptcy in the event of project failure.

The failure of a project can seldom be tolerated by the project promoters, as it could induce corporate or national bankruptcy. These projects exhibit a dichotomy of risk by having to attract financial investment and avoid any prospect of failure and yet involving large-scale, long-term construction and future operations in a volatile world market to return the non-recourse loans. Therefore risk analyses must commence at pre-feasibility stage and form a continuous and integrated part of the project planning.

The difference in public and private sector projects is again evident. Private sector projects have to meet strict criteria, and weaknesses at any point usually results in abandonment. However, while public sector projects often have a strong financial base they also introduce the concept of worthwhileness. Worthwhileness can be assessed from either a cost–benefit analysis or a subjective points evaluation system, which might include political desirability, and other comparative rather than absolute project parameters.

14.4 UK offshore projects

For the large steel platform projects the most typical strategy has been separate design contracts in series for the main structure and for the topsides, separate fabrication contracts in parallel for the substructure and the topsides units, and a separate contract for the subsequent sea transport of all the above and its installation offshore. The conditions of contract have been developed from those used for large industrial projects onshore. For the very large integral gravity platforms the UK and Norwegian promoters have placed one comprehensive contract for design and construction of the substructure, and separate contracts for systems, equipment, topsides module fabrication, etc. as above. For pipelines the promoters have placed separate contracts for design, fabrication, laying and connecting in.

For smaller projects the tendency has been to continue the onshore practice for industrial projects of a comprehensive contract, but of course differing from onshore projects in that construction is in the contractor's and subcontractors' yards up to load-out for installation offshore.

Reimbursable terms of payment were common in the contracts for design, fabrication, installation, hook-up and commissioning for the first projects, but have tended to be replaced with fixed-price terms plus unit rates where appropriate, giving competition on price without the loss of flexibility to cope with uncertainty.

14.5 Partnering

To compete successfully for international work requires knowledge of conditions, working practices, cultural differences and climatic conditions. Many of the most potentially profitable contracts are for complex and expensive projects, which also have the greatest financial and technical risk for the contractor. An increasing number of countries insist that local firms, are employed on all contracts. Hence many UK engineering firms have formed temporary or semi-permanent joint ventures or consortia with local firms, or have appointed local promoters. In addition, arrangements have been made with other UK or EU or international firms to acquire the necessary expertise and to share the risk of the project. The majority of this competition is not based on the UK system, and experience of fixed-price contracting had to be acquired.

Industrial and multidisciplinary projects are often long-term, expensive and complex undertakings. It is in no party's interest to incur additional costs and/or time delays because of the adversarial organizational structure and contract strategy historically adopted by that industry sector. Consequently, in many countries the concept of partnering has become increasingly attractive.

Partnering: a long-term commitment between two or more organizations for the purpose of achieving specific business objectives by maximizing the effectiveness of each participant's resources. The relationship is based on trust, dedication to common goals, and an understanding of each other's individual expectations and values (non-contractual) (CII).

Alliance partnership: joint management of a contract between promoter and one or more contractors.

In any project for both promoters and contractors, the main reasons for partnering are:

'To attain mutually desirable goals, to satisfy long-term needs, and to achieve future competitive advantage ... to achieve mutual profit and mutual success for both companies.'

'To achieve specific business objectives by maximizing the effectiveness of each participant's resources. The relationship is based on trust, dedication to common goals, and an understanding of each other's individual expectations and values.' (CII.)

Until recently most practitioners considered that *term partnering* – that is, partnering arrangements made over a long period of time and usually for a number of different projects – was the only effective method. The principal reasons given were that it was felt that the relationship had to be long-standing to secure benefit and that culture change could not be effected in a single project. More recently it has been recognized that the objective of continuous improvement is impossible to achieve if the relationship becomes too easy.

Partnering – either term or project specific (PSP) – requires a culture change from all parties concerned with the project. The adversarial relationship has to be completely replaced, and this will require revision or innovation in project management techniques and in project staff training and development. Periodic critique – a self-evaluation methodology using a form completed by all stakeholders – is essential if the partnering arrangement is to be continuously improved.

All parties need to be responsive to the needs of their partners. For example, if the contractor needs a drawing in less than the period allowed by contract then the architect or consultant engineer needs to accommodate this. The focus must be on funding timely cost-effective solutions to problems, and that demands input from all, regardless of who is responsible for direct implementation. Letter writing is not the answer, open communication and trust through partnering is.

Alternative dispute resolution (ADR)

Particularly in the USA, where litigation can account for between 8 and 12% of project costs, the contract strategy of partnering has been inherently linked with the formal application of ADR techniques. There needs to be agreement on alternate disputes resolution methods for use in these cases – facilitated negotiation, arbitration, mini-trial, disputes review board etc. – as litigation is not the way to resolve most issues.

As partnering requires parties to be responsive to the needs of others, and being responsive requires commitment of resources, cost-effective organizations need to accept some risk based on the trust that the other party will do the right things for the right reasons.

There has been some concern that partnering might contravene the spirit if not the letter of the EU Public Procurement Directives. To comply with these requirements in a PSP arrangement the project manager for the promoter would have to advertise in the *Official Journal* the intention to prepare a preselected list of bidders, and describe the selection criteria. The factors to be taken into consideration in contract award would need to be clearly shown in the invitation documentation.

14.6 International market for projects

In 1988 the world market for building (residential, commercial and industrial), civil engineering and process engineering was $94 100bn. From the 1970s the bulk of the new projects were in the Middle Eastern countries based on the oil economies with surplus capital due to the quadrupling of the price of oil. By 1980 the Middle East market was down to about a third and large increases in work in Africa and the Americas had occurred. Into the 1990s the major market was in the USA and Canada, which accounted for about two thirds of the international market.

For the UK in 1992 the Single European Market (SEM), which permits the free movement of goods, people, services and capital, produced a combined internal construction market seven times larger than the current UK domestic market. Competition is regulated by strict procedures imposed on the procurement of major projects to ensure free and fair competition within the market, and ultimately the goal is to harmonize the procurement of projects throughout the SEM. However, this is likely to take some time, given the different legal systems amongst the 12 countries together with the addition of Austria, Sweden and Finland in 1995 and the association of several countries from eastern Europe in the near future.

Many countries in Central and Eastern Europe – in particular the more developed countries of Russia, the Czech Republic, Hungary and Poland – have a policy of restructuring a state economy into a market economy, which is difficult to implement but as markets become established there appears to be tremendous potential if funding can be arranged. Some financial assistance is being made available from the European Bank for Reconstruction and Development, and the Japanese Overseas Development Administration has awarded 1000bn yen over ten years to help Japanese firms become established in this market. The possibilities of adopting private finance contracts for international tender, hence avoiding hard currency problems, are also being investigated.

There is also an area of high economic growth closer to Japan consisting of the newly industrialised countries,, as classified by the Organization for Economic Cooperation and Development, and some Asian countries, including South Korea, Taiwan, Singapore, Hong Kong, Thailand, Indonesia, the Philippines, Malaysia and Brunei. This region is forecast to produce growth over the next ten years about two and a half times greater than the rate forecast for the USA. Owing to their proximity to this market, Japanese companies have a turnover about three times greater than the combined turnover of the Western industrialized countries. However, the UK, with its long-established connections with Hong Kong, Singapore, Malaysia and Brunei, has established a presence in this market.

14.7 Management of culture

The management of culture has not yet become the major component of project management development that may ultimately be needed. This need has been partially avoided by companies by tending to undertake work for promoter companies wholly or partially owned by organizations from the same country of origin, or by working largely in countries with a background of colonial, commonwealth or trading relations.

A feature of the management of culture is the process of localization. This introduces a blend of personnel from the host country at all levels in the organization, and matches the management style more closely to the domestic practice. A variety of skills can be provided to assist with this problem, and training programmes in languages, overseas postgraduate study in project management and business subject areas can be of assistance. Education coupled with experience of international projects is viewed as the basis for developing a new approach.

The UK will need to invest in project management education, in on-the-job training and in research and development to meet these challenges. The current company structures are well suited to this task, but a greater investment in professional project management personnel is likely to be required. The UK will need to retain competitive advantage in its domestic market, compete in other EU countries' markets and become competitive in new markets. It is always difficult to predict the future and as has been demonstrated international markets change with economic performance.

Further reading

Benitez Codas, M.M. (1992) Cultural integration in bi-national joint ventures. In: *Project Management without Boundaries*, 11th Internet World Congress on Project Management, Florence, pp. 155–164.

Bussey, L.E. (1978) *The Economic Analysis of Industrial Projects*, Prentice-Hall, Englewood Cliffs, NJ.

Cleland, D.I. and Gareis, R. (eds) (1994) *Global Project Management*, McGraw-Hill, New York.

Construction Industry Institute (1990). Interim Report on Partnering, CII, Austin, TX.

Merett, A.J. and Sykes, A. (1973) *The Finance and Analysis of Capital Projects*, 2nd edn, Longman, London.

NEDO (1991) *Partnering: Contracting Without Conflict*, National Economic Development Office, London.

Schwartz, E.A. (1986) Disputes between joint ventures: a case study. *International Construction Law Review*, **3** (4) p. 360.

Chapter 15
Turnkey and BOOT Projects

The turnkey type of contract is widely used for the delivery of projects. The basic concept of this approach is for the organization requiring the project to contract with a single organization that would be responsible for design, procurement, engineering and commissioning. Literally all the promoter would have to do would be to 'turn a key in the door' and the project would be operational. This chapter first describes the procedures relating to the use of turnkey contracts in project management. In conclusion, a variation of the turnkey contract, the concession project, often referred to as BOOT project, is discussed. In this form of contract the contractor effectively becomes the promoter and, in addition to the role of turnkey contractor, also finances the project and operates and maintains the project over a period of time to generate sufficient income to provide a commercial return.

15.1 Comparison of turnkey contracts and BOOT projects

Concession projects, often referred to as BOOT projects, may be considered as contractor-financed turnkey contracts with the operation and maintenance element extended to generate sufficient revenue to service the debt on the financial investment and sufficient profit over that period before transferring the facility to the 'principal'.

As many turnkey contractors have been responsible for operating and maintaining facilities for a number of years after commissioning, this has resulted in many turnkey contractors gaining all the necessary expertise to take the role of promoter in concession projects.

Concession project strategies grew out of turnkey contracting, in which the supplier of complicated plant would remain in charge as an operator for a defined period of time to train personnel and prove that projects could meet warranted performance specification and capacities. Under the turnkey arrangement, operation elements evolved into longer operational periods under which the supplier of plant and equipment realized a more substantial proportion of his reward from the actual operation and maintenance of the facility.

In conventional turnkey contracts, governments have attempted to shift the risk for the project construction to the private sector while still bearing the risk of financing and operating the project. In a concession project the major risks of finance and operation, however, are borne by the promoter.

In turnkey contracts feasibility studies are often carried out by the principal to determine the basic requirements of a facility. An invitation to tender often includes a *performance specification* on which the contractor will base his bid. In concession projects a promoter would be responsible for feasibility studies in a speculative bid and the principal for an invited bid. Commercialization of a turnkey contract is normally the responsibility of the principal, who will often pay the contractor a mobilization fee and monthly payments for the work carried out. In a typical concession project, however, commercialization will be carried out by the promoter.

Construction of a turnkey contract will be carried out by the contractor who will in most cases operate it for a period of up to two years after commissioning, the principal then taking over the operation for the life of the facility. In concession projects the promoter will often enter into a construction contract with a contractor and operate the facility over the concession period before finally transferring the facility to the principal.

A comparison of turnkey contracts and BOOT projects and the major risk allocation is illustrated in Table 15.1.

Table 15.1 Comparison of major risk allocation between turnkey contracts and BOOT projects.

Project phase	Project type	
	Turnkey	BOOT
Feasibility	Principal	Principal/promoter
Commercialize	Principal	Promoter
Construction	Contractor	Promoter
Operation	Principal	Promoter

15.2 Definition of turnkey contracts

A turnkey contract may be defined as follows.

> Turnkey preparation of a facility means that a single contractor acquires and sets up all necessary premises, equipment, and supplies operating personnel to bring a project to a state of operational readiness. All the customer needs to do is turn the key to begin full effective usage of the new facility. Sometimes the contractor continues to operate the facility for the customer; in other cases the customer assumes operational control. Turnkey facilities are appropriate for customers who are unable to perform or wish to avoid their own subcontracting for ordering and testing components acquired from several vendors. Recruiting, screening and training is a highly specialized task. A turnkey contractor is compensated either through surcharges on each item or service procured for the facility or by a commitment in advance to a fixed price.

15.3 Performance specification

A turnkey performance specification forms the basis of the contract between a promoter and a contractor. This specification indicates the required performance of a facility, often producing a defined offtake for a specific period of time. This specification will often form the basis of a contractor's design, construction and commissioning methods and operation and maintenance of a facility to meet a promoter's requirements. Many turnkey contractors will often only require a performance specification, standards and conditions of contract from a promoter's organization to prepare a turnkey tender.

The performance specification for a particular project may be determined by a promoter organization, by a consultant advising the promoter, or by a promoter and contractor, depending on the type of contract preferred.

A turnkey contract is often adopted on the basis of the findings of a feasibility study carried out by or on behalf of a promoter organization. A promoter's choice of plant and type of contract and the clarity of the performance specification adopted will be paramount in achieving an efficient plant. The contractor's design obligation, irrespective of the specification adopted, will form part of his general obligation to supply a facility that meets the required performance specification and guarantees. In turnkey contracts it is normal to provide only proven technology to meet the requirements of the performance specification.

The content of information issued to a contractor may be dependent on the results of a feasibility study and a number of possible design methods proposed by the promoter or his representative. Drawings and site investigation reports may be provided to the contractor, but in many turnkey contracts the contractor will be responsible for ensuring that ground conditions are sufficient and for the preparation of working contract drawings. A promoter may require either a schedule of rates or values to be used to assess interim payments and projected budget figures and manufacturers' drawings for stage payments on design and manufacture of plant and machinery. The most common types of turnkey contract are usually related to specialized or patented process plant supply contractors with the advantage of obviating split responsibility and interface problems.

Various standard forms of conditions of contract have been obtainable in the UK for many years to cover civil engineering, mechanical and electrical works. The type of contract adopted would be determined by the promoter as best suited to meet his needs, with a number of different contract procedures currently available. A project may be considered as a single contract, in which case a main contractor will sign a contract with the promoter and accept full responsibility for the proper execution of the works. The main contractor may be a large civil engineering contractor who will employ specialist subcontractors but purchase plant from a plant manufacturer. If the main contractor is a process plant contractor he may engage a subcontractor for the civil engineering works and for the electrical installations.

Many traditional forms of contract strategy necessitate the production of detailed specifications, formal contract documents, bills of quantities and detailed working drawings, which will be very time-consuming but will ensure that bids can be evaluated on a similar information base. Promoters pursuing this traditional form of contract may also nominate subcontractors or suppliers, who will need to be coordinated by the promoter. Should the promoter prefer a turnkey contract strategy then he may require the contractor to design, construct, inspect, test, commission and hand over a completed facility ready for operation. Such contracts will usually be entered into for projects that involve some specialist process, often proprietary equipment, in which only a few large, integrated organizations are truly competent.

15.4 Economic choice

The choice of a turnkey contract is often determined on the basis of funds available to a promoter as well as technical reasons. Export financing has tended to favour the turnkey contract, and many developing countries have been eager to adopt a turnkey contract as a means of financing a project. The lump sum price offered in a turnkey contract is often considered an advantage by promoters who have a limited budget and are not in a position to incur additional costs, as the price is determined at an early stage in the evaluation process.

When promoters are providing their own finance for a project then technical expertise and speed of construction are often the main reasons for the choice of a turnkey contract. A promoter needs to identify the economic reasons for the choice of a turnkey contract compared with other traditional forms of contract. The promoter costs associated with planning, designing, supervising and coordinating a contract are often significant, and often a managing contractor would be best employed on major high-technology contracts where a turnkey contract would not be in a promoter's best interest. The need to carry out capital cost estimates, equipment selection and economic criteria in regard of a number of proposals is of prime importance to a promoter organization.

A promoter will need to investigate a project's cash flow, operation and maintenance methods, and the payment method during the contract period. The fact that a turnkey contract is considered to have a shorter overall duration from conception to commissioning may be a major criterion in a promoter's choice. The cash flow often based on the contractor's proposed payment schedule may also affect the choice of a turnkey contract. Many turnkey contracts are agreed on a lump sum basis and contain a list of unit prices for certain sections of the works. These prices often serve for the valuation of variations and interim payments.

In a turnkey contract the promoter may consider that the best terms of payment are on a schedule of values and not on the basis of monthly measurement of quantities. The method of payment based on agreed schedules may also be linked to the construction programme, which will allow the promoter to identify each interim payment by simply monitoring progress.

15.5 Bidding turnkey contracts

Many promoter organizations with in-house engineering capabilities often consider a turnkey contract to be in the promoter's best interest for a particular project. If the need for a short tender and construction duration is paramount, the promoter may choose to adopt a turnkey contract in which tenderers are invited to submit an offer for the design and construction of a particular project, which may include commissioning and hand-over and possibly operation and maintenance and training for a specified duration. A turnkey contract provides for all elements of the project up to and including the commissioning and hand-over of the plant as an operational unit. Process design, civil engineering and structural engineering design are all included.

Normally a contractor's offer will be in the form of a lump sum, with or without a price fluctuation clause, depending on the terms of contract, the type, location and size of the project, the duration of contract and many other factors. Consequently it is usually possible to obtain tenders fairly quickly, particularly when compared with the time required to prepare detailed designs and specifications and invite tenders in the more traditional manner.

As a bidding contractor is usually permitted a degree of freedom in selecting a process design, he can adopt a design suited to his own particular expertise, specialization and choice. If a promoter decides to approach a number of turnkey contractors indicating performance requirements, then the promoter will receive a number of different proposals and costs, allowing a number of options to be examined. The options would give the promoter far more flexibility over the traditional type of contract based on one design. In most cases turnkey contractors use past experience and feedback techniques to ensure that current designs are up to date and suitable to the promoter's requirements.

The turnkey lump sum also requires the most precise definition of the project objectives prior to a contract price, and introduces an adversarial relationship between the owner and the turnkey contractor, who must give consideration to economic survival as well as to the best interests of the promoter. A promoter's need for urgency may result in the choice of a turnkey contract. If urgency is a major criterion then the turnkey contract is considered most suitable as the turnkey contractor is well accustomed to the degree of urgency normally attached to such tenders. Also, the difficulties normally associated with subcontractors and project coordination and the burden on the promoter's resources are significantly reduced.

It is, however, extremely important that bidders are given a comprehensive brief, otherwise it will be extremely difficult, if not impossible, to compare accurately the bids received. The more comprehensive the brief, the more comparable will be the tenders. Many unsuccessful turnkey projects have often resulted from an inadequate definition of requirements at the tender stage. One method considered for turnkey tender submission is that the tender should be in two parts, one part (the technical package) containing the product specification documents and the other (the financial package) containing the price documents.

The technical package contains components regarding the construction and operation of the facility, often determined by the performance specification, and the financial package and its components are determined by the financial requirements of the contract. A typical schedule of components that may be included in a turnkey contract is as follows.

Technical package: standards and specifications, quality of offtake operation and maintenance, plant, equipment and material schedules, construction and labour programmes, method statements, preliminary calculations, designs and drawings, performance guarantees and procurement schedules.

Financial package: selling price of the facility, schedule of values, mobilization fee, financial guarantees, bonding arrangements, projected drawdown of payments, payment milestones, insurances and ratio of host country and overseas currency expenditure.

The selling price is normally fixed and based on a schedule of values for construction, commissioning and operation and maintenance.

15.6 Projects realized utilizing turnkey contracts

The turnkey approach is quite frequently adopted for process/industrial type projects. The types of project carried out on a turnkey basis both in the UK and overseas are considerable. Plants have been and are still being constructed under turnkey contracts, examples of which include a crude oil supply line in Algeria, a power station for the Manx Electricity Authority and the Jing-Aw Hilton Hotel complex in Shanghai. These contracts have all required multidisciplinary involvement with short construction durations.

Turnkey contracts are often considered when a project requires a multidisciplinary involvement by promoter organizations who consider that type of contract best suited to their requirements or do not have the resources available to coordinate and supervise the project. Total responsibility and accountability for all aspects of the project in the hands of a single turnkey contractor often determines a promoter's choice in multidisciplinary projects. Some promoters, however, prefer to retain in-house project management teams when they consider a project too large and requiring resources not available to a turnkey contractor.

The location of a plant may well determine the choice of a turnkey contract where little or no expertise is available to carry out a multidisciplinary project. An existing plant constructed under a turnkey contract, which has proved to be efficient and is similar in its process requirements to one proposed by a promoter, may well determine the final choice, but as in most turnkey situations a fixed design will result in a less flexible system. The use of turnkey contracts has also been considered for the provision of trunk roads and in the UK a small number of design and build contracts for road schemes have been awarded. In multidisciplinary turnkey contracts, however, the contractor carries the responsibility for all aspects of the contract from conception through to commissioning and often into operating the plant and training permanent staff for the promoter.

15.7 Advantages and disadvantages of turnkey contracts

Promoter organizations are in a position to compare all the economic advantages and disadvantages of the turnkey contract for individual projects.

The advantages and disadvantages of turnkey contracts are considered below.

Advantages

- In projects undertaken by governments or state-owned enterprises, ownership and control after contract completion is retained in the hands of the owner. This is especially true in the case of the traditional turnkey contract, where the involvement of the turnkey contractor could be eliminated once the contract is completed, as the contractor would have no share in capital ownership, and there would be no conflict in policies and management of the operations of the enterprise.
- In the turnkey contract a major advantage to the promoter stems from the fact that the responsibility for the contract lies with a single source, and the promoter is relieved from responsibilities for the equipment or plant and performance.
- The turnkey contract generally ensures that the project is put into operation more rapidly than other contracts as both design and construction are the responsibility of one entity.
- When a turnkey contract extends beyond the commissioning stage

the teething problems associated with a multidisciplinary project can be resolved by the contractor's trained personnel.

Disadvantages

- The cost of a turnkey contract may be significantly higher than a traditional form of contract because cost estimates are often expressed in overall terms without a detailed breakdown.
- The turnkey contract does not allow enough participation of the promoter or familiarization with the facility that the promoter will operate after hand-over.
- The turnkey contract does not permit the normal checking procedure associated with a traditional form of contract.
- The turnkey contract should not be adopted when domestic technological services are sought to be developed by the promoter.

The turnkey system may also bring together companies in joint ventures who on an individual basis would not normally have tendered. Although joint ventures may offer many advantages, disputes may often arise between the partners regarding the contract responsibilities of each partner, with adverse effect on the project undertaken. Amongst the disadvantages is the inclination of a plant manufacturer to include a maximum of plant of his own manufacture. Other problems stem from differences between the contract conditions included in the offer and the promoter's requirements. Unfortunately these often arise because the brief and tender conditions are not clearly defined.

Generally the advantages of both design and build and turnkey contracts are that the promoter need only deal with one organization; there are no disputed responsibilities; a firm price can be obtained; and time can be shortened by overlapping the later stages of design with the early stages of construction. Because all the work is carried out by a single contractor effective quality control can be provided without the need for redundant reviews.

15.8 Concession contracts

There has been a growing trend in recent years both in the UK and overseas for principals, usually governments or their agencies, to place major projects into the private sector rather than the traditional domain of the public sector by using concession or BOOT project strategies. The adoption of this form of contract strategy has led a number of organizations to consider its implementation for different types of facilities, on both a domestic and international basis and by speculative or invited offers.

Privatized infrastructure can be traced back to the eighteenth century, when a concession contract was granted to provide drinking water to the city of Paris. During the nineteenth century ambitious projects such as the Suez Canal and Trans-Siberian Railway were constructed, financed and owned by private companies under concession contracts.

The transfer element of a BOOT project implies that after a specified time the facility is transferred to the principal; this cannot be considered as real privatization. In a BOO project, however, ownership of the facility is retained by the promoter for as long as desired, and this is therefore more consistent with the concept of privatization.

In the late 1970s and early 1980s some of the major international contracting companies and a number of developing countries began to explore the possibilities of promoting privately owned and operated infrastructure projects financed on a non-recourse basis under a concession contract.

The term BOT was introduced in the early 1980s by the Turkish Prime Minister Turgat Ozal to designate a 'build, own and transfer' or a 'build, operate and transfer' project; this term is often referred to as the *Ozal Formula*.

15.9 Definition of BOOT projects

A build-own-operate-transfer (BOOT) project, sometimes referred to as a concession contract, may be defined as:

> a project based on the granting of a concession by a principal, usually a government, to a promoter, sometimes known as the concessionaire, who is responsible for the construction, financing, operation and maintenance of a facility over the period of the concession before finally transferring the facility, at no cost to the principal, as a fully operational facility. During the concession period the promoter owns and operates the facility and collects revenues in order to repay the financing and investment costs, maintain and operate the facility and make a margin of profit (Smith and Merna, 1992).

Other acronyms used to describe concession contracts include:

FBOOT	finance, build, own, operate, transfer
BOO	build, own, operate

BOL	build, operate, lease
DBOM	design, build, operate, maintain
DBOT	design, build, operate, transfer
BOD	build, operate, deliver
BOOST	build, own, operate, subsidies, transfer
BRT	build, rent, transfer
BTO	build, transfer, operate
BOT	build, operate, transfer

Many of these terms are alternative names for BOOT projects, but some denote projects that differ from the above definition in one or more particular aspects, but which have broadly adopted the main functions of the BOOT strategy.

15.10 Organizational and contractual structure

A typical BOOT structure illustrating the number of organizations and contractual arrangements that may be required to realize a particular project is shown in Figure 15.1.

The key organizations and contracts include the following.

Principal: responsible for granting a concession and the ultimate owner of the facility after transfer. Principals are often governments, government agencies or regulated monopolies. The structured contract between the principal and promoter is known as the *concession agreement*. It is the document that identifies and allocates the risks associated with the construction, operation, maintenance, finance and revenue packages and the terms of the concession relating to a facility. The preparation and evaluation of a BOOT project bid is based on the terms and project conditions of the *structured concession agreement* (SCA).

Promoter: the organization that is granted the concession to build, own, operate and transfer a facility. Promoter organizations are often construction companies or operators or joint venture organizations incorporating constructors, operators, suppliers, vendors, lenders and shareholders.

The following organizations and contracts may be included within the BOOT project strategy:

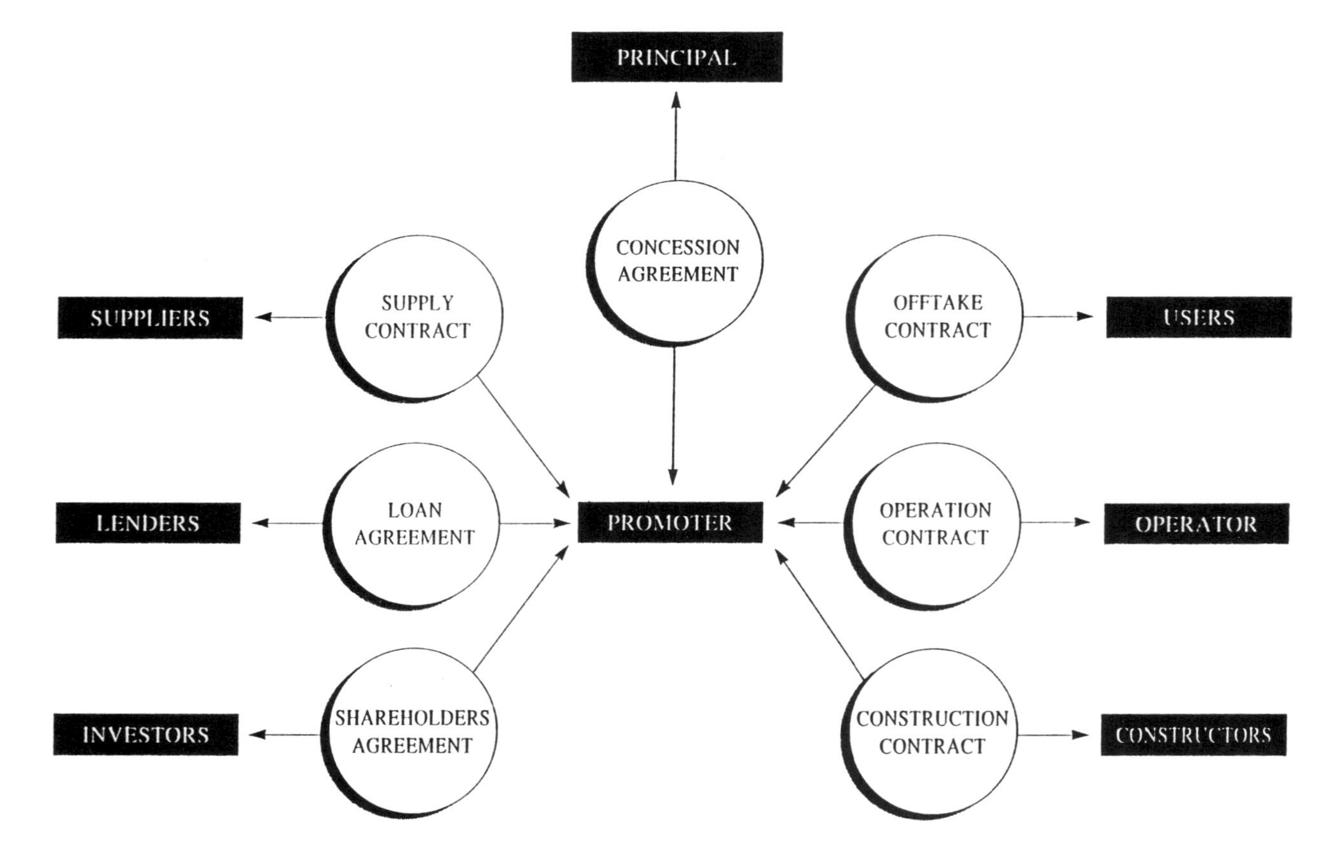

Figure 15.1 Organizational and contractual structure for BOOT.

Supply contract: contract between the supplier and promoter. Suppliers are often a state-owned agency, a private company or a regulated monopoly who supply raw materials to the facility during the operation period.

Offtake contract: in contract-led projects such as power generation plants a 'sales' or 'offtake contract' is often entered into between the user and the promoter. Users are the organizations or individuals purchasing the offtake or using the facility itself. In market-led projects, however, such as toll roads or estuarial crossings, where revenues are generated on the basis of directly payable tolls for the use of a facility, an offtake contract is not usually possible.

Loan agreement: the basis of the contract between the lender and the promoter. Lenders are often commercial banks, niche banks, pension funds or export credit agencies who provide the loans in the form of debt to finance a particular facility. In most cases one lender will take the lead role for a lending consortium or a number of syndicated loans.

Operations contract: contract between the operator and the promoter. Operators are often drawn from specialist operation companies or companies created specifically for the operation and maintenance of one particular facility.

Shareholder agreement: contract between investors and the promoter. Investors purchase equity or provide goods in kind and form part of the corporate structure. These may include suppliers, vendors, constructors and operators and major financial institutions as well as private individual shareholders. Investors provide equity to finance the facility, the amount often determined by the debt/equity ratio required by lenders or the concession agreement.

Construction contract: contract between the constructor and the promoter. Constructors are often drawn from individual construction companies or a joint venture of specialist construction companies. Constructors can sometimes take, and have taken the role of promoters for a number of BOOT projects both in the UK and overseas.

15.11 Concession agreements

In BOOT contracts a concession agreement is used as the basis of the contract. The concession agreement forms the contract between the principal and the promoter, and is the document that identifies and allocates the risks associated with the construction, operation and maintenance, finance and revenue packages and the terms of the concession relating to a facility over the lifetime of the concession before transfer of the facility to the principal.

The statutory concession agreement is adopted when governments are required to ratify a treaty that may lead to legislation and consequent concessions. The terms of a concession granted under statute may usually only be altered or varied by the enactment of further legislation. Under this form of agreement the promoter (concessionaire) would be required to enforce his rights by making an application to the courts for judicial review of a principal's (government's) actions. In the case of concessions granted under statute, third parties may potentially have the right to apply to the courts to enforce provisions of the concession that the government do not wish to apply. A statutory concession agreement was adopted for the Channel Fixed Link project.

The contractual concession agreement is often adopted when one government organization enters into an agreement with a promoter to undertake a specific concession. In the contractual agreement either party may amend or relax the terms of the agreement. Under such an agreement any breach of the concession by the principal would entitle the promoter to damages, and in some cases specific performance of the terms of the concession. In the contractual concession only the parties to the concession can enforce the terms. A contractual concession agreement was adopted for the North–South Expressway in Malaysia.

In some cases the concession agreement is a hybrid form of both contractual and statutory elements. This form of agreement is often adopted when a principal or promoter requires an element of legislative control and the benefits associated with the contractual agreement. If for example planning consent is considered a major requirement for implementation of the concession then a statutory element may be incorporated into the hybrid form to cover this requirement.

15.12 Procurement strategies

BOOT projects are procured by invited tender from the principal or by a speculative bid from a promoter group to an individual principal. In the case of an invited bid many elements of the risk may be determined by the 'terms of invitation'. In the case of a speculative bid the promoter will need to approach the principal to determine his obligations under the terms of the concession agreement. Worldwide, about 60% of BOOT projects are awarded as the result of speculative tendering, but within the EU this percentage is much lower.

Speculative bids

A speculative bid is one by which a promoter approaches a principal with a proposed scheme considered commercially viable by the promoter and requests the principal to grant a concession to the promoter to build, own and operate a facility for a defined period of time before transferring the facility to the principal. A speculative bid is for a concession usually undertaken by the principal and requires the promoter to prepare a concession agreement as the basis of the bid. Many BOOT projects begin when promoters approach governments privately to propose a much-needed project that government finds difficult to finance from the public sector budget.

In projects involving new transport infrastructure projects a speculative bid is considered as one in which the private sector promotes an innovative project from concept to meet a perceived market need.

Invited bids

An invited bid is one by which the principal invites a number of promoters to bid in competition for the privilege of being granted a concession to operate a facility normally undertaken by the principal or one of his organizations. An invited bid is often a solicitation of bids based on a speculative proposal made open to tender.

In the case of invited bids for transport infrastructure projects a route will have been through a public inquiry, and the public sector seeks the support of the private sector to design, build, finance and operate a route. Alternatively, the government may have defined the transport corridor and requires the private sector to select the actual route, assemble the land, design, build, own and operate the project on the basis of a time-limited concession. In the UK the practical consequence of privately financed projects requires each project to be authorized individually by an Act of Parliament. In invited bids such as the Dartford River Crossing a hybrid bill procedure was adopted to authorize the project's go-ahead, and the invitation involved:

- the identification by the government of a corridor for the proposed route;
- a competition for the financing, building and operation of a road serving the corridor, inviting bids from private companies;
- the promotion by the government of a hybrid bill to authorize the road, the tolls, land acquisition and arrangements for the concession.

Competitive bids for BOOT projects should follow the normal procedure for awarding public works projects; ideally the government would identify the project and define the project specifications, nature of government support, the proposed method of calculating the toll or tariff, the required debt/equity ratio and other parameters for the transaction.

The government would then invite preliminary proposals, with the winner selected on the basis of normal competitive criteria such as price, experience, track record of the promoter, or on the basis of side benefits to the host country.

15.13 Concession periods

Concession periods are sometimes referred to as a lease from government. The concession period normally includes the construction period as well as the operation and maintenance period before transfer to the principal. However, in the case of the Shajiao Power Plant constructed in China, the concession period of 10 years excluded the construction time. In the Macau Water Supply Project a concession period of 35 years included the refurbishment of existing plants rather than construction of new facilities but also permitted the principal to repurchase the rights if deemed necessary. This agreement also permits the concession to be extended by mutual agreement of the parties, which in effect constitutes an addendum to the contract. A provision for an extension of the concession period, in this case the operating period, may be included in the concession agreement to protect a promoter against the principal defaulting on his contractual obligations, which may result in projected returns not being met. Typical concession periods range from 10 to 55 years when granted by governments under a BOOT initiative.

In infrastructure projects the concession period is often longer than for industrial facilities. Concession periods of between 10 to 15 years may be successfully financed, but principals need to accept that the project economics must be strong enough to bear the enhanced depreciation rate over such short periods as well as the return required on the capital investment. In the Dartford Crossing project the concession period was a maximum of 20 years, which could be terminated at no cost to the principal as soon as surplus funds had been accrued by the promoter to service all outstanding debt.

The concession period should be sufficient to:

- allow the promoter to recover his investment and make sufficient profit within that period to make the project worthwhile;
- but not allow the promoter to overcharge users when in a monopolistic position having already recovered his investment and made sufficient profit.

The commercial viability of a concession contract and the difficulty and uncertainty of predicting revenues over long periods of time is a major obstacle to many promoter organizations. If there is flexibility in the concession agreement to adjust the concession period then this may reduce the promoter's risk and allow predicted revenues to be achieved.

15.14 Existing facilities

In a number of concession contracts an existing facility is included as part of the concession offered or requested. This may be a requirement of the principal in an invited bid, or offered as an incentive by a promoter in a speculative bid. In some industrial/process facilities, however, a promoter may only agree to tender a particular project provided he is given operational control of an existing facility which may affect the performance of the facility to be tendered.

The operation of an existing facility often guarantees the promoter an immediate income, which may reduce loans and repay lenders and investors early on in the project cycle. The commercial success or failure of an existing facility must be considered by the promoter at bidding stage in order to determine the success or failure of the proposed concession.

Principals can influence pricing mechanisms by making available to promoters existing facilities that are capable of earning revenues during the construction period. In the Sydney Harbour Tunnel project, revenues generated by the existing bridge crossing are shared between the principal and the promoter, which enables the promoter to generate income to service part of the debt prior to completion of the tunnel.

Assets capable of producing earnings that can be used to pay capital costs, debt service and operating expenses are a familiar feature of BOOT projects.

A concession to operate a section of existing highway generated $16m for the promoter of the North–South Expressway in Malaysia. The concession to operate existing tunnels as part of the Dartford Bridge Crossing project offered an existing cash flow but required the promoter to accept the existing debt on those tunnels. An existing concession also formed part of the concession agreement for the Bangkok Expressway; this arrangement required the operation and maintenance of an existing toll highway, with generated revenues being shared between the principal and the promoter.

15.15 Classification of BOOT projects

BOOT projects may be classified on the basis of the method of procurement, the type of facility, the location of the facility and the method of revenue generation.

Speculative–invited

In a speculative bid the promoter will determine which costs and risks should be borne by his own organization and which should be borne by the principal and other parties involved. In an invited bid the principal will determine the concession and the costs and risks to be borne by the promoter under the terms of invitation.

Infrastructure–industrial/process

The components of each of the packages of an infrastructure project will contain costs and risk levels that are different from those considered for an industrial or process plant. An infrastructure project may require large capital expenditure during construction but operate on a small budget. A process or industrial facility, however, may require a low capital expenditure but require a high operating budget over the operation phase.

The number and types of risks associated with a BOOT project may often be determined by the type of facility and the number of contracts and agreements to be included.

Infrastructure facilities may often be considered as static or dynamic facilities. A road or fixed bridge facility may be considered as static, as the facility offers no moving parts and requires no input of raw materials or power; usually the static facility will have a smaller operation and maintenance cost than a dynamic facility. A light transit railway facility will require a major power source during the operation phase, which will result in operational costs being higher than for a static facility.

Domestic–international

The location of a project will determine the host country's political, legal and commercial requirements, which will be a major factor in project sanction. In a domestic project the promoter will often be aware of the country requirements and have access to local financial markets. In international projects promoters may need to carry out in-country surveys to determine risks associated with meeting the requirements of the concession and determine how revenues may be repatriated to service loans. In effect each international project will be determined by the constraints of the host country government.

Market-led–contract-led

One of the major risk areas associated with BOOT projects is the generation of revenues, which often leads to market-led revenues being far more uncertain than those based on predetermined sales contracts. The commercial risks to be considered in a BOOT project are often determined by the revenue classification.

The demand for a toll road that depends solely on revenues from users may be much lower than forecasted at feasibility stage, this may be due to increased costs of fuel or a reluctance by users to pay tolls. In the case of a water treatment plant revenues will be contract-led, and provided demand is met and an effective price variation formula takes inflation into account a promoter may consider the risk associated with revenue negligible compared with other risks.

In a number of market-led projects promoter organizations will often seek contract-led revenue streams to reduce the risks associated with revenue generation. In a toll road facility promoter organizations may approach haulage contractors and enter into take-or-pay agreements for the use of the facility; in effect the promoter organization is providing lending organizations with a guaranteed source of revenue, which will reduce the risks associated with the finance package.

In summary, the number of organizations, contracts, data and resources required to meet the project and the major risk areas will be determined by the classification of the project.

15.16 Projects suitable for BOOT

BOOT projects are a means of meeting the needs associated with population growth, such as housing, water sanitation and transportation; industrial growth such as power, infrastructure and fixed investments; tourism and recreation such as airports, hotels and resorts; and environmental concerns such as waste incineration and pollution control. Developing nations are receptive to the idea of funding such projects under a BOOT strategy, which will often reduce the capital and operating costs and reduce the risks normally borne by the principal. Provided sufficient demand exists for these projects, revenue streams can be identified and the commercial viability determined by promoters and lenders.

The two most fundamental constraints on project development are economics and finance. In a BOOT project the promoter must cover operating expenses, interest and amortization of loans and returns on equity from project revenues. However, promoters often consider the suitability of a project based on global market forces and the commercial viability of a project, which affect the profitability rather than the facility itself.

Any public service facility that has the capacity to generate revenues through charging a tariff on throughput may be considered suitable for a BOOT strategy provided suitable financing can be achieved. The most successful BOOT projects will be those in the small to medium range – up to US$500m – as private sector equity requirements for such projects are usually obtainable.

Tolled highways, bridges and tunnels, water, gas or oil pipelines and hydroelectric facilities are considered suitable projects, as a private economic equilibrium is obtainable. However, subsidies are often necessary for high-speed train networks and light rail trains, as prices paid by users are often low and governments generally prefer to control prices.

The characteristics of BOOT projects are particularly appropriate for infrastructure development projects such as toll roads, mass transit railways and power generation, and as such they have a political dimension of public good that does not occur in other privately financed projects.

15.17 Risks fundamental to BOOT projects

The two types of risk fundamental to BOOT projects are elemental risks and global risks. Elemental risks are defined as those risks that are contained in the elements of the project; global risks are defined as those risks outside the elements of the project that nevertheless influence the concession agreement.

Many of the global risks are addressed and allocated through the concession agreement, with elemental risks retained either by the promoter or allocated through the construction, operation and finance contracts. The author has developed a structured concession agreement based on the terms of the concession and the project conditions specifically for BOOT contracts. This structured concession agreement is used as the basis for risk analysis, bid preparation and bid evaluation.

There are two phases when risks associated with financing BOOT projects occur:

- the pre-completion phase relative to construction risks (the construction phase);
- the post-completion phase relative to operational risks, with the first few years of operation being the major operation risk (the operation phase).

Promoters are exposed to risks throughout the life of the project, which may be summarized as:

- failure at several stages of the project;
- failure in the later stages of the project when considerable amounts of money have been expended in development costs;
- failure of the project to generate returns, without the opportunity to recover costs.

Risks associated with market prices, financing, technology, revenue collection and political issues are major factors in BOOT projects. Risks encountered on BOOT projects may also include physical risks such as damage to work in progress, damage to plant and equipment and injury to third persons and theoretical risks such as contractual obligations, delays, *force majeure*, revenue loss and financial guarantees.

The major risk elements of a BOOT project may be summarized as:

Completion risk:	the risk that the project will be not completed on time or to budget.
Performance and operating risk:	the risk that the project will not perform as expected.
Cash flow risk:	the risk of interruptions or changes to the project cash flow.
Inflation and foreign exchange risk:	the risk that inflation and foreign exchange rates effect the project costs and revenues.
Insurable risks:	risks associated with equipment and plant (commercially insurable risks).
Uninsurable risks:	*force majeure.*
Political risk:	risks associated with sovereign risk and breach by the principal of specific undertakings provided in the concession agreement.
Commercial risk:	risks associated with demand and market forces.

Demand risks associated with infrastructure projects are much greater than those for facilities producing a product offtake, as an infrastructure project is static and cannot normally find another market whereas a product may be sold to a number of offtakers through the life of the concession. Facilities producing an offtake bear the risk of product obsolescence and competition, which usually leads to market risks dominating, especially when operation and maintenance costs are high and concession periods short.

15.18 BOOT package structure

The structure of a BOOT project is highly sophisticated, requiring the full participation of all the parties involved in identifying and allocating the relevant project risks and responsibilities and an appreciation of the political, legal, commercial, social and environmental considerations that have to be taken into account when preparing BOOT project submissions. The process of developing a BOOT project is immensely complicated, time-consuming and expensive.

The major components of a BOOT project include:

Build: design, manage project implementation, carry out procurement, construct and finance.

Own: own the asset for the concession period and the licence for the equipment used.

Operate: manage and operate plant, carry out maintenance, deliver product or service and receive offtake payment.

Transfer: hand over plant in operating condition at the end of the concession period.

The number of components and their timing over the concession period need to be identified at an early stage of a project. This should be addressed in a format that can be utilized to identify the obligations and risks of each organization involved in the project so that an equitable risk allocation may be determined.

BOOT contracts may be determined by four major packages:

- *Construction package.* Containing all the components associated with building a facility, normally undertaken in the pre-completion phase and may include: feasibility studies, site investigation, design, construction, supervision, land purchase, commissioning, procurement, insurances and legal contracts.
- *Operational package.* Containing all the components associated with operating and where applicable owning the facility and may include: operation, maintenance, training, offtake, supply, transfer, consumables, insurances, guarantees, warranties, licences and power contracts.
- *Financial package.* Containing all the components associated with financing the building, and in some cases the early stages of operation and may include: debt finance loan, equity finance loan, standby loan agreements, shareholder agreements, currency contracts and debt service arrangements.
- *Revenue package.* Containing all the components associated with revenue generation and may include: demand data, toll or tariff levels, assignment of revenues, toll or tariff structures and revenues from associated developments.

This structure incorporates all the components of a BOOT contract into discrete packages over both the pre-completion and the post-completion phase of the concession period. The type of facility, its location and revenue realization would effectively be contained in one of the packages. Having identified and allocated components into the four packages a promoter organization could then determine the risk associated with each package and how such risks would be shared. The package structure provides a rational basis for financial appraisal of BOOT contracts, for the allocation of risk within the concession agreement and contractually between the parties concerned, and for the structure of the tendering process.

15.19 Advantages and disadvantages of BOOT projects

The BOOT project may offer both direct and indirect advantages for developing countries:

- promotion of private investment
- completion of projects on time without cost overruns
- good management and efficient operation
- transfer of new and advanced technology
- utilization of foreign companies' resources
- new foreign capital injections into the economy
- additional financial source for priority projects
- no inroads on public debt
- no burden on public budget for infrastructure development
- positive effect on the credibility of the host country.

The introduction of new technologies, project design and implementation and management techniques are considered as advantageous to developing countries; the disadvantages however are host country constraints and financial market constraints.

A major advantage of a BOOT project is the financial advantage to a government, as its off balance sheet impact does not appear as a sovereign debt.

The advantages to an overseas principal are that he does not need to compete for scarce foreign exchange from the state purse, there is a specific need for the project, and risks are transferred to the promoter. The most important attractions to governments of Asian developing countries is off balance sheet financing, transfer of risk, speedy implementation and an acceptable face of privatization.

The involvement of the private sector and the presence of market forces in BOOT schemes ensure that only projects of financial value are considered.

There are arguments for and against BOOT projects. The arguments for are as follows.

- *Additionality.* This would offer the possibility of realizing a project that would otherwise not be built
- *Credibility.* This would propose that the willingness of equity investors and lenders to accept the risks would indicate the project was commercially viable
- *Efficiencies.* The promoter's control and continuing economic interest in design, construction and operation of a project will produce significant cost efficiencies, which will benefit the host country
- *Benchmark.* The usefulness to the host government to use a BOOT project as a benchmark to measure the efficiency of a similar public sector project
- *Technology transfer and training.* The continued direct involvement of the project company would promote a continuous transfer of technology, which would ultimately be passed on to the host country. A strong training programme would leave a fully trained local staff at the end of the concession period
- *Privatization.* A BOOT project will have obvious appeal to a government seeking to move its local economy into the private sector.

The arguments against:

- *Additionality.* Commercial lenders and export credit guarantee agencies will be constrained by the same host country risks whether or not the BOOT approach is adopted
- *Credibility.* This benefit may be lost if the host government provides too much support for a BOOT project, resulting in the promoter bearing no real risk
- *Complication.* A BOOT project is a highly complicated cost structure, which requires time, money, patience and sophistication to negotiate and bring to fruition. The overall cost to a host government is greater than that of traditional public sector projects, although proponents of the BOOT approach argue that overall costs are less when design and operating efficiencies are taken into account and compared with public sector alternatives.

Although there are a number of advantages and benefits associated with BOOT projects very few BOOT proposals have reached the construction stage. A review of BOOT schemes by an EU Commission concluded that there were three key problems associated with BOOT projects: availability of experienced developers and equity investors; the ability of governments to provide the necessary support; and the workability of corporate and financial structures.

The risks associated with BOOT projects are far greater than those considered under traditional forms of contract as the revenues generated by the operational facility must be sufficient to pay for construction, operation and maintenance and finance. The uncertainty of demand and hence revenues, cost of finance, length of concession periods, levels of tolls and tariffs, effects of commercial, political, legal and environmental factors are only some of the risks to be considered by promoter organizations.

Further reading

Augenblick, M. and Custer, S.B. (1990) *The build, operate and transfer (BOT) approach to infrastructure projects in developing countries.* Working papers – Infrastructure, The World Bank.

Nicklisch, F. (1990) Performance guarantees in turnkey heavy plant contracts. *International Construction Law Review*, **7** (2), 250–256.

Renault, J. (1989) Perspective of the contractor as the investor. In: *Reprints of Second International Construction Projects Conference.*

Road Transport Research (1987) *Toll Financing and Private Sector Involvement in Road Infrastructure Development*, Organization for Economic Cooperation and Development, Paris.

Smith, N.J. and Merna, A. (1992) Investment appraisal and risk management for BOOT contracts. In: *Project Management Without Boundaries*, 11th INTERNET World Congress of Project Management, Florence, Italy, vol. 2, pp. 191–199. Edizioni Unicopli SpA, Florence.

Svenska Konsultforeningen (1971) Turnkey polemic. *Consulting Engineer*, December, 45–47.

Chapter 16

Contractual Arrangements in the EU

Contracts and contractual procedures are inherently related to the project management approach to any given project. As described in Chapter 10 the contract is a mechanism for the equitable allocation of risk and the transfer of work and motivation. This chapter considers some of the more important differences in contractual arrangements in some EU countries and the implications for practical project management.

16.1 Development of the European Union

The number, the size, the strategic alliances and the relative influence of European countries changed and continued to change throughout the eighteenth and nineteenth centuries. By 1900 a recognizable structure had been established, although considerable adjustment took place as a result of the two World Wars and a division of Europe into 'Western' and 'Eastern' spheres of influence. This latter change is only now being addressed with the restructuring of many central and eastern European countries from command economies to market economies and the formulation of an EU policy to admit eastern European members from the turn of the century.

The current EU member states are all countries from the Western European group, although it could be argued that the members joining in 1994, particularly the Scandinavian countries of Sweden and Finland, marked the beginnings of an expanded European organization.

The formation of the EU was largely based on the international organizations and charters to put it into practice, largely with the backing of the USA, at the end of 1945 onwards. In 1946 at Bretton Woods the International Monetary Fund and the World Bank were formed to help re-establish the European economies; this was coupled with the launching of the integrated programme, the 'Marshall Plan', to relieve poverty and encourage industry (Male and Stocks 1991).

These developments provided the incentive for the formation of the Organization for European Economic Cooperation (OEEC) in 1948, and the General Agreement on Tariffs and Trade (GATT) was signed. This provided for customs duties on certain items to be harmonized; it was operated by more than 20 countries, and demonstrated the principles for trade within the future Single European Market.

Belgium, Holland and Luxembourg (Benelux) decided to cooperate more closely on trade and tariffs by having no charges on internal trade and common rates for all external trade. This was followed in 1950 by the French 'Schuman Plan', which was a proposal that the French and German production of coal and steel should be controlled by a single authority. This developed into the Treaty of Paris, 1951, when France, Germany, Italy, Holland, Belgium and Luxembourg agreed to form the European Coal and Steel Community (ECSC), which was the first really supra-national European organization.

In 1957 two further treaties were signed, creating the European Economic Community (EEC) and the European Community of Atomic Energy (EURATOM). The member states of the EEC were the same six countries that had formed the ECSC six years earlier. The EEC was the most important of the three communities, and was primarily responsible for the economic integration that provided the basis for the European Union.

The original six members have admitted new member states between 1973 and 1994. The first Treaty of Accession was drafted in 1972 as a result of discussions with the UK, Denmark, Eire and Norway. Norway decided against joining, but the other three were admitted in 1973. Greece followed in 1981, and in 1986 Spain and Portugal were admitted to bring the EC up to 12 member states. In 1982 it was agreed that completion of the Single Internal Market should be achieved by the end of 1992, and member states began to ratify the necessary directives to remove all barriers to the free movement of goods, services, capital and people. In 1993 the EU came into being. More recently, the expansion of members has continued with the acceptance of three further countries for admission in 1994: Austria, Finland and Sweden.

The EU provides a single market of over 350 million people, comparable to the US market of around 420 million people. There will still be major differences between each of the member states for the foreseeable future, but harmonization and integration of practices in the management of engineering and technological projects is already beginning and is likely to become increasingly significant. Work on continuing integration was formalized with the drawing up of the Treaty of Maastricht which, with certain exceptions negotiated by the UK and Denmark, provides for closer cooperation and the establishment of a central bank and single currency. The treaty has to be renegotiated and ratified by the member states in 1996 and target dates for key events have been established by the Councils of Ministers: for example, the current target date for a single currency is 1998.

16.2 UK practice

Contract practice and systems of management vary in the UK between industrial, process, civil engineering, building construction, and offshore projects. Within each of these sectors there is also great variety of practice. However, UK practice relies on the use of sets of 'model' conditions for contracts for construction, published by government departments, professional institutions and trade associations. These models consist of terms of contract that specify the responsibilities of the parties and the formal system of communications between them, as described in Chapter 12. For most projects the public and private promoters add to these models or use their own equivalents.

Over the last 40 years British industry has maintained the same basic structure. The public sector promoters have initiated a range of transportation and energy infrastructure projects although increasingly this function is being devolved into the regulated private sector. Private sector promoters have initiated the majority of commercial, residential, industrial and process projects. The contracting companies range from multinational contractors to small, specialist firms, and the UK has a large number of relatively small consulting engineering partnerships.

For over 150 years UK consulting engineers have operated as partnerships. A partnership has unlimited liability. Many consultants are engaged on large and complex projects, and the threat of financial ruin is raised by the increasing numbers of promoters, contractors and members of the public who resort to litigation to resolve disputes. To overcome these problems a number of consulting partnerships have become part of larger multidisciplinary consulting firms, while others have merged to form larger limited liability companies.

Most contracts are price based: admeasurement or lump sum. The admeasurement contract does not guarantee the promoter the final price. The permanent works are remeasured, and payment is based upon the final quantities for each of the priced BoQ items. Hence the BoQ fulfils two functions: it has to have a low total price to be successful in competition and it also has to maximize the financial return to the contractor by careful pricing. Variations to the works and claims made under the contract which also affect the final price are an accepted part of the UK industry. This has given UK contractors a reputation for disputes and financial overruns with European promoters, who are more accustomed to fixed-price lump sum contracting.

Contracting companies are relatively low-cost, high-risk businesses, and during the recent recession in the UK economy many of these companies became attractive targets for acquisition or merger. In 1987 over 2000 UK companies were acquired, with a market value in excess of £28bn. Acquisition is basically a choice between development or modification of an in-house facility and the purchase of an already operating facility. There are capital costs and operational costs associated with acquisitions, and for an acquisition to be successful there must be the possibility of a significant improvement in economic performance. This trend has resulted in many contractors' becoming part of large national or multinational corporate organizations owned by UK, other EU and non-EU parent companies.

These organizational and business changes have resulted in changes to the conventional contract strategy. The conventional system is seen as not placing sufficient emphasis on time and cost performance. There is also pressure for forms of contract strategy that facilitate buildability, reduce the risk to the promoter, allow promoter involvement in management, and permit early starts. Two increasingly popular strategies are the management contract and the package deal.

16.3 Legal systems in the EU countries

The reports and other publications used show that the terms of contracts in the countries covered vary for the obvious reasons of differences in legal systems and in the extent that liabilities and relationships in public works contracts are established by law.

As would be expected, the public sector projects are subject to more national regulation than the commercial projects. There are substantial differences in legal regulation between the six EU countries described here. Regulation is least in the UK.

There also appear to be differences from country to country in the extent that national and EU law is enforced, but legal regulation is not obviously the main cause of differences in contract arrangements. Problems due to the law governing contracts and construction were never mentioned to us when obtaining the information on the case studies.

The design and construction of public works projects in all the countries normally have to conform to standard specifications and codes, but the extent to which legislation requires the certification of design, public liability insurance, registration of engineers, contractors, craftsmen and others or regulates tendering or the resolution of disputes varies. In none does the law govern the important choices of deciding the number and scope of contracts. Nor does the law dictate how contracts are managed. Legal advice may be needed in preparing and managing a contract, but the primary task for promoters in any country is to define what project they want, state their priorities between quality, time and cost, assess what are the risks, foresee what may change, and then choose terms of contract (so far as permitted by any mandatory legal rules) that are most likely to control any consequential problems.

16.4 Projects in EU countries

The engineering industries in the EU countries are directly affected in terms of design standards, procurement, the use of products and the free and fair competition for EU public works contracts.

Procurement methods vary between the Member States, but there are three standard methods of tendering for construction work that are used in the majority of Member States: open tendering, selective tendering, and negotiation. Sometimes particular works require specialized skills, and promoters have adopted the package deal, or management contract. Unlike the UK, many EU Member States do not tender on the basis of precise quantities, which means that a great deal of emphasis is placed on the preparation and clarity of the written specification.

Terms of contracts vary between building, infrastructure and industrial projects in many EU countries, and they vary with the size of projects. One model set of conditions of contract dominates construction in one EU country, Germany. The variety of models used appears to be greatest in the UK, and so does the practice of modifying them project by project. Table 16.1 indicates some of the more commonly used model forms.

Practice varies most from country to country in their contracts for constructing public buildings, and varies least for the large industrial projects, but practice in each sector is not uniform.

A summary of some of the key differences within the EU is given below. These notes highlight the main countries undertaking large industrial projects and hence includes comments on contractual practice in France, Germany, Italy, the Netherlands and Spain.

Table 16.1 Model conditions of contract.

European

General Conditions for the Supply of Plant and Machinery for Export, United Nations Economic Commission for Europe (ECE) (alternatives for supply with and without erection).

General Conditions for Work Overseas, 2nd edition, 1990, European Development Fund.

International

Conditions of Contract (International) for Works of Civil Engineering Construction, 4th edition, 1987, revised 1988, Fédération Internationale des Ingénieurs Conseils (FIDIC). Known as the FIDIC 'Red Book'.

Conditions of Contract (International) for Electrical and Mechanical Works, 3rd edition, 1987, Fédération Internationale des Ingénieurs Conseils (FIDIC). Known as the FIDIC 'Yellow Book'.

Model forms of contract for turnkey contracts for the construction of fertilizer plants, 1981, UNIDO (with guidelines published 1982).

Projects in France

Three main tendering procedures are currently in use:

Adjudication is conventional open tendering which is adopted for simple supply contracts only.

Appel d'offre is the most common procurement method. It is based upon competitive tendering with open or restricted tender lists, against a detailed specification. The work is offered to the most attractive bidder in terms of the overall cost, the technical value and the likely duration.

Public works contracts are usually announced by public notice and are tendered competitively, with restrictions for the major contracts. A deposit (*cautionnement*) of 5% has to be paid. The contract documents usually include a specification, conditions of contract, drawings and sometimes a BoQ. When tenders are submitted they are usually accompanied by variations, as alternatives to tender documentation are encouraged. The tender bids are opened in the presence of the tenderers and the lowest bid is accepted; the tenderer has 24 hours to accept, and may withdraw if a payment equal to the difference to the next lowest bid is made.

For the construction of industrial projects, including oil, chemical and nuclear projects, the differences from UK practice are as follows.

- The choice of subcontractors and vendors is sometimes subject to consent by the promoter.
- Fixed-price stage payments are made to the contractor, less retention. Subcontracts often paid by unit rates. By law a subcontractor may be due payment direct from the *maître d'ouvrage*.
- Variations and their costs are negotiated between promoter and contractor.
- Conditions for comprehensive contracts: promoters use their own sets, particularly the international oil and chemicals manufacturers, but the contracts are in effect similar to the UK IChemE models except that there is no role of the engineer.

Industrial projects in Germany

The eight local authorities, *Länder*, and the two city states administer the public sector contracts usually under the *Verdingungsordnung für Bauleistungen* (VOB) regulations. The VOB are in three sections: Part A deals with the method of tendering; Part B with the general conditions of contract, the specification and other documents; and Part C includes the 49 DIN standards and notes on measurement. The majority of public works are procured by advertisement and open competitive tender, whereas private sector work uses the method of invitation and negotiation. The most common form of contract is the fixed-price, lump sum contract based upon a specification and approximate BoQ. The award is not necessarily made to the lowest. Winning tenders are evaluated on the basis of a price to performance ratio but details of how this assessment will be calculated are not provided to the contractor. In certain circumstances restricted tendering is used.

For the construction of industrial projects, including oil, chemical and nuclear projects, the differences from UK practice are as follows.

- The contractor is responsible for detailed design and construction to achieve performance, including subcontractors' and vendors' items unless otherwise negotiated, but subject to supervision by the promoter. The larger promoters usually specify design standards. Smaller promoters place turnkey contracts.
- Design is usually led and managed by the contractor's staff, but often supplemented by design by consultants and by vendors.
- There is one main contract for a project. Some large projects are undertaken by joint ventures or consortia or are split into several contracts to suit the engineering capacity of contractors.

- The choice of subcontractors/vendors may be subject to consent by the promoter.
- Fixed-price stage payments are made to the contractor, less retention unless the contractor's bank has provided a performance guarantee.
- Variations and their costs are negotiated between promoter and contractor.
- The contractor can be entitled for extra time for a few reasons.
- Conditions for comprehensive contracts: some promoters use their own sets of terms, particularly the international oil and chemicals manufacturers, but these are in effect similar to the UK IChemE models except that there is no role of the engineer in contracts.

Industrial projects in Italy

There is a high degree of regulation of the range of public works contracts:

asta publica is an open tendering procedure in which the contract is awarded to the lowest bidder.

licitazione privata is the closest method to UK practice; bidders have to undergo a prequalification process, and generally the lowest-priced bid is awarded the contract, although delivery, operating cost and technical merit are all assessed in the evaluation procedure.

appalto concorso is similar but usually contains detailed specifications and is used for complex projects. A proportion of work can be awarded by negotiated tender.

trattativa privata requires no advertising or competition.

For the construction of industrial projects, including oil, chemical and nuclear projects for the state-owned promoters, the differences from UK practice are as follows.

- The contractor is responsible for detailed design and construction to achieve performance, including subcontractors' and vendors' items unless otherwise negotiated, but subject to supervision by the promoter. The larger promoters usually specify design standards.
- Design is usually led and managed by the contractor's staff, but often supplemented by design by consultants and by vendors.
- There is extensive subcontracting/purchasing from vendors. The choice of subcontractors/vendors is subject to consent by the promoter.

- Variations and their costs are negotiated between promoter and contractor.
- Conditions of comprehensive contracts: promoters use their own sets of terms, particularly the international oil and chemicals manufacturers, but these are in effect similar to the UK IChemE models except that there is no role of the engineer.

Industrial projects in the Netherlands

In the Netherlands most private sector and public sector engineering contracts use the Ministry of Public Works contract, in conjunction with a BoQ. There is no special administrative law applying to contracts with the state. The *Rijkswaterstaat* tender regulations mean that work is awarded to the most acceptable tenderer. A small proportion of work is awarded by open tendering or negotiated tendering, but the majority of work is awarded on the basis of selected tendering. In addition, each commissioning body has its own regulations. Returned, competitive tender bids are opened in public, and contracts are awarded to the tenderer whose bid appears to be the most acceptable and not necessarily the lowest. However, in about 98% of cases the contract is awarded to the lowest-priced bid. There has been some privatization in recent years, and this has encouraged the use of design and build contracts and negotiated tendering procedures.

Industrial projects, including oil and chemical projects, are usually split into two phases: basic design, including process design and producing the project specification and statement of requirements; and detailed design, procurement and construction.

Basic design may be by the promoter or a design contractor, to a fixed lump sum price or on man-hour rates. Contractors are selected on knowledge of them and their experience. Many contracts are negotiated.

The detailed design is usually part of an 'EPC services' (engineer-procure-construction management) contract, but subject to supervision by the promoter or the design contractor. The larger promoters usually specify design standards.

Contractors bid competitively, but when urgent the contract may be negotiated with the basic design contractor. Bank guarantees may be required from contractors with their bids.

The EPC services contractor places contracts for equipment, materials, construction and services on behalf of the promoter.

Depending on the extent of definition of scope and the flexibility required, payment may be as above, but is more often lump sum than in the first phase. It may also include incentives for achieving cost, time, quality and safety criteria. Payment is usually monthly, based upon progress, less retention. Variations are paid for only if caused by reasons beyond the contractor's control, based upon unit prices and/or hourly rates. Disputes are rare, but those there are, are usually settled by arbitration.

Alternatively a turnkey contract (paid lump sum fixed price) is used, if the detail is sufficient and time allows, but this is rare for large projects.

Industrial projects in Spain

The contract arrangements vary greatly, depending upon the promoter's experience and the complexity, urgency and novelty of projects. Many are similar to UK practice but without the use of nationally recognized conditions of contract.

Most major contracts are awarded to general contractors, who subsequently employ subcontractors as required. The public works contracts are subject to regulation, but the basic principle is that contractors compete for work on the basis of the lowest acceptable tender figure:

subasta is a system of open, competitive tendering. The contractor has to calculate the detailed quantities and submit a price below the client's estimate.

concurso is a method of qualified tendering under which the record, the experience, the staff and the technical capabilities of the firm are considered in addition to the tender price. This approach is often used for design and construction work.

subasta con admision previa is a form of selected competitive tendering whereby the tender list is narrowed down prior to the submission of bids. The return bids are then assessed on the basis of lowest price.

contraction directa is a form of negotiated tender, which is used for a minority of contracts.

Conditions of comprehensive contracts: promoters use their own sets of terms, particularly the international oil and chemicals manufacturers, but these are in effect similar to the UK IChemE models except that there is no role of the engineer.

16.5 Harmonization within the EU

Harmonization is the process of removing the physical, technical and fiscal barriers by adjusting the community law, taxes, markets and procedures of the Member States to conform to agreed community standards. It is estimated that harmonization should produce a 5–6% increase in internal trade, which would be worth about US$250 000m. In the short time between the publication of the Act and the creation of the Single European Market, from 1986 to 1992, it was not possible to complete all the necessary arrangements. Therefore the main objective of harmonization is not to remove all national diversity, except when the diversity is perceived as a barrier to the four freedoms of the Single European Market.

Engineering and technology projects are primarily affected by four key Directives, dealing with public works contracts, compliance, construction products, and public supply contracts. However, significant penetration of non-domestic markets requires the additional harmonization of the design techniques, insurance and liability and professional qualifications. Ultimately national design codes will be replaced by Eurocodes.

EU legislation already in force is intended to regulate the selection of contractors for commercial and public projects. Whether the trend towards EU harmonization will ultimately result in the standardization of conditions of contract is at present uncertain, but it is clear that there will be distinct national differences in contractual practice and procedure for the foreseeable future.

16.6 Innovation

Promoters, the government and others in the UK have become increasingly interested and active in innovations in contract arrangements, probably motivated by recurrent experience of the costs of poor quality, late completion and contractual disputes, and the uncertainties of predicting whether these will occur. Innovations have followed the privatization of what were public services in the UK, because of the change towards commercial rather than political accountability. The harmonization of EU practice provides a vehicle for potential innovation, which could be beneficial for project managers.

Experienced private and public promoters and many contractors have stated that simpler engineering and construction contracts should be used in the UK. Conditions of contract have tended to become longer and longer, and have increased rather than decreased the promoters' potential risks because terms in them have become impractical and the documents too complex for project managers or others to comprehend how to apply them. Additions made for one project become a precedent for the next, and further additions are made contract by contract that increase contractors' liabilities, contrary to agreed principles for allocating risks. The main argument has been that it would be more cost-efficient for promoters and contractors collectively to agree on what are the essential general terms and end the practice of adding 'supplementary conditions' for each project to meet what may be illusions about its special needs.

Obtaining successful results also depends upon prior attention to relationships with contractors so as to anticipate problems that might lead to conflict. The basis of non-adversarial contracts is not new, but some promoters appear to be unfamiliar with them and the corporate attitudes they need to be successful. It is the promoter who must therefore decide the priorities, allocate the risks, propose the terms of contract, select the contractor and manage the uncertainties.

In order to utilize unfamiliar types of contract a promoter usually requires additional managerial and contractual expertise. This requirement is often met by the appointment of an internal or external project manager. To be successful in using innovative contracts the project manager needs the authority to represent all interests in the promoter's organization.

Further reading

Male, S.P. and Stocks, R.K. (1991) *Competitive Advantage in Europe*, Butterworth-Heinemann, Oxford.

Smith, N.J. and Wearne, S.H. (1993) *Construction Contract Arrangements in EU Countries*, European Construction Institute, TF003/4.

Stammers, J.R. (1992) *Civil Engineering in Europe*, McGraw-Hill, Maidenhead.

Chapter 17

Project Management in Developing Countries

Civil engineering and the construction industry in developing countries are sufficiently different to warrant the inclusion of this chapter in a book on project management. The range of types and size of construction companies is different, the environment in which they operate is different, the resources that are employed may be different, and the way projects are funded is different. This chapter reviews some of the main issues that contribute to the distinctive nature of developing countries and how these affect projects.

17.1 Differences in developing countries

Construction projects, and the construction industry, in developing countries are significantly different from those in the developed industrialized world. The main differences are related to climate, population and human resources, material resources, finance and economics, and socio-cultural factors. Due recognition of these differences is a prerequisite for the successful management of projects in developing countries.

Climate

Many poor developing countries experience quite different climatic conditions from those in the temperate North. The type of project that is required, the most appropriate technology to be applied, and the way in which the project is managed can be influenced by these variations in climatic conditions. For example, communities living in hot climates have quite different requirements for power and water, giving rise to alternative approaches to the planning and design of the requisite infrastructure facilities. Climate will also affect the design and type of technology used: solar power may be a realistic alternative to thermal power generation; high temperatures and long hours of sunlight may indicate alternative forms of sewage treatment such as waste stabilization ponds; and the design of buildings must be aimed at reducing glare from sunlight and ensuring that heat is kept out (rather than in).

During construction, it may be necessary to take precautions not required in cooler climates, such as chilling or adding crushed ice to the water used in mixing concrete, and paying particular attention to the curing of concrete. Planning and scheduling of construction work can also be affected by the climate – particularly when constructing roads, bridges, and hydraulic structures in areas affected by heavy seasonal monsoons. The project manager therefore needs to be fully aware of the climatic implications from the very earliest stages of the project.

Population and human resources

Of the nearly six billion people living on this planet, less than one billion live in what the World Bank categorizes as high-income countries, while some 3.5 billion live in low-income countries, two billion of these in China and India. Up to 25% of children born in many low-income countries do not reach the age of five, and the average life expectancy in the poorest countries is less than 40 years. Over a billion people do not have access to a safe water supply, a third of the world's population have inadequate sanitation facilities; and more than two million deaths per year are directly attributable to polluted water. In addition to this lack of basic water and sanitation facilities, many live in substandard housing; transportation and communication links are poor or non-existent; and fuel for heating and cooking is in short supply.

The problem is compounded by exponential population growth – the current estimates are for a world population of seven or eight billion by the end of the decade. In the past, much of the developing world's population has been in rural areas, but this is now changing, and population in urban areas is growing faster than in the rural areas. In 1990 there was a worldwide urban population of over 2.5 billion, of which some 850 million were living in large cities: that is, those having a population in excess of one million. Of these 850 million, over 600 million are in developing countries. By the year 2000, it is likely that this world urban population will rise to around 3.5 billion, and that almost half the population of the developing world will live in urban areas.

These facts and figures point to a huge need for infrastructure development throughout the developing world – a need that can only be fulfilled through the implementation of well-managed construction projects.

Population and human resources not only affect the need for projects but also the way projects are implemented. The large pool of available and relatively cheap labour, much of which is unemployed or under-employed, points to a less mechanized approach to construction and a greater use of human labour. Labour-intensive construction requires a different approach to the planning, design and management of projects, and these issues must be addressed at the earliest stages of the project. Although the labour force in developing countries may be plentiful, it is also likely to be relatively unskilled. The questions of training and technology transfer therefore need to be taken into consideration throughout the planning and implementation of the project.

Materials, equipment and plant

Many of the materials commonly used in construction projects are often not readily available in developing countries. Cement and steel may have to be imported and paid for with scarce foreign exchange. Delays in importation and difficulties in gaining passage for imported goods through customs are not uncommon and need to be allowed for. Even if materials are manufactured in the country of the project, supplies cannot always be guaranteed, and the quality may be inferior to that normally expected in industrialized countries. Production capacity and quality should therefore be assessed before detailed design is done.

In some cases, it may be necessary to consider alternative materials, such as stabilized soil, ferrocement, round pole timber, or pozzolana as a cement replacement. Many such alternatives are traditional indigenous building materials and may be more acceptable than steel or concrete. An assessment of what is available and appropriate needs to be made at an early stage and used in the design.

Imported mechanical equipment, whether it be for construction or for incorporation within the completed project, is expensive and requires maintenance. Trained maintenance technicians and a reliable supply of spare parts are the absolute minimum requirements if the equipment is to continue to function over its anticipated life. There is generally a shortage of reliable and operable construction plant in developing countries, and it is often not possible to hire plant, because plant hire companies do not exist. Managers of projects, particularly large ones, will have to import plant for the project, and then decide on whether to sell it or transport it back to the home country on completion of the project.

Finance and economics

Although there is a great need for new projects in developing countries, there is also a lack of funds from the normal sources expected in the developed countries. Many projects are funded externally from national aid agencies, international development banks, or non-governmental organizations (NGOs) such as international charities. Project managers involved in the identification, preparation, and appraisal stages of a funded project need to be fully aware of the requirements of the grant- or loan-awarding agency to whom they are making application for funding, as each has its own specific requirements.

Socio-cultural factors

The successful management of a project in a developing country requires an understanding of the ways society is organized and the indigenous cultural and religious traditions. In Muslim countries time must be allowed for workers to participate in daily prayers and, during the month of Ramadan, fasting is mandatory during daylight hours, thus affecting productivity and the way work is organized. The respective roles of men, women, religious and community leaders, and landowners must be understood, particularly when managing projects in which the community is actively participating.

If socio-cultural factors are not taken into consideration, a project may not be successful even if it is successfully constructed. A new water supply may not be used if the community feel they do not own it, or if traditional existing sources of water have a strong cultural significance; sanitation facilities might be under-used or neglected if the orientation offends religious beliefs or if men's and women's toilet blocks are sited too close together. As with other factors already mentioned, a knowledge of socio-cultural influences is therefore necessary at the earliest stages of a project, because they may have a significant effect on project identification, appraisal and design, as well as on construction and operation.

Working with other professionals

Project managers often have to work with a variety of professionals – electrical and mechanical engineers, chemical engineers, heating and ventilating engineers, environmental scientists, architects, quantity surveyors, and planners for example. But nowhere is the diversity of professionals so great as when the project is located in a developing country. In addition to the list above, the project manager working in a developing country may well have to work closely with agricultural scientists, health and community workers, educationalists and professional trainers, economists, community leaders, sociologists, ecologists, epidemiologists, local and national politicians, and perhaps many others. Good written and oral communication skills and an ability to understand the views and perspectives of other professions are valuable qualities in any project manager, but they could be vital when the project is set in a developing country.

17.2 The construction industry in developing countries

Unlike the developed countries, many developing countries do not have a mature construction industry consisting of well-established contracting and consulting companies. Much, if not most, of the building and construction is done by the informal sector. This consists of individual builders and tradesmen, who are mainly concerned with building family shelters, and community and self-help groups, who may construct small irrigation works, community buildings, grain storage facilities, water wells and the like. These individuals and small groups rarely attract funding, and the works are completed using labour-intensive methods and locally available materials. Small community groups may gain the attention of national and international non-governmental organizations, who are more inclined to fund and work with such groups than are formal aid agencies and development banks.

The formal sector consists of public or state-owned organizations and private domestic contractors. The proportion of work carried out by the public sector is usually much higher in the less developed countries than it is in the richer countries, but low salaries, lack of incentives, and poor promotion prospects often result in highly demotivated professional and technical personnel.

Contracting is a risky business in any country, but in many poor developing countries the lack of access to financing, excessively complex contract documents, failure to ensure fair procurement practices, the high cost of importing equipment, and the fluctuations of demand for construction often mean that the private sector of the construction industry has not had the opportunity to establish itself sufficiently to bid for major infrastructure projects. These are almost entirely carried out by international contractors and funded by national and international loans and grants. In some countries, up to 80% of major building and civil engineering is executed in this way.

Because of the importance of the construction industry to the development of the overall economy, the growth of an indigenous construction industry needs to be encouraged. In an attempt to do this, the World Bank and other agencies have encouraged the 'slicing' and 'packaging' of contracts – breaking up large contracts into smaller ones that local contractors would have the resources and capability to bid for. Other initiatives have included a greater use of labour-intensive construction, help with finance for smaller construction companies, the encouragement of plant hire companies, and management training for principals and staff of construction companies. Technology transfer and training, particularly management training, play an important role in this.

17.3 Finance and funding

Developing countries are by definition poor. Funding for projects will therefore be scarce, loan finance difficult to obtain, and resources scarce. In many cases the only source of finance will be from development banks, aid agencies, or charitable non-governmental agencies, many of whom obtain at least part of their funding from national aid agencies.

The major development banks include the World Bank, the Asian Development Bank (ADB), the African Development Bank (AfDB), and the European Bank for Reconstruction and Development (EBRD). These are all multilateral funding agencies, drawing their funds from several different countries. They operate as commercial banks, lending money at agreed rates of interest. The loans have to be repaid, albeit the loan conditions are often more favourable than commercial banks, and they may allow a period of grace before repayments commence.

Most industrialized countries have their own government bilateral aid agencies, such as the UK's Overseas Development Administration (ODA). These agencies fund projects in developing countries through loans and grants, and also direct some of their allocated funds to those development banks of which they are members. Aid awarded directly by these agencies is often 'tied': that is, the grant or loan is conditional upon some of the goods and services needed for the project being procured from the donor country.

Loan finance for construction companies to expand, buy equipment, or simply to maintain adequate balances and ease cash flow difficulties, is extremely difficult to obtain from commercial banks in developing countries. Contractors may therefore be forced to borrow from other sources at inflated rates of interest. However, some countries have development finance companies, which act as intermediaries to channel funding from external agencies such as the World Bank to the construction industry and other developers.

17.4 Appropriate technology

The distinctive nature of the construction industry in developing countries suggests alternative approaches to the design, construction, and management of projects. The application of appropriate technology is one approach that has been promoted as a way to overcome some of the problems associated with the implementation and long-term sustainability of development projects in the Third World. Appropriate technology should be able to satisfy the requirements for fitness for purpose in the particular environment in which it is to be used. It should also be maintainable using local resources, and it should be affordable. Many would argue that all technology should be appropriate, and perhaps therefore it is intermediate technology that we should be focusing our attention upon.

The concept of intermediate technology was first developed by E.F. Schumacher, who defined it in terms of the equipment cost per workplace. He suggested that the traditional indigenous technology of the Third World could be represented as a £1 technology, while that of the industrialized world was a £1000 technology. An example, from the agricultural sector, is the traditional hand or garden hoe as a £1 technology, compared with a modern tractor and plough as the £1000 technology. Schumacher pointed out that, throughout the world, the equipment cost of a workplace was approximately equal to the average annual income, and that any budding entrepreneur could save sufficient money over a 10–15 year period to purchase the equipment necessary to start a small business. However, if the budding entrepreneur is a resident of a developing country, where salaries are a fraction of that in the industrialized world, it would take him or her over 100 years to purchase advanced equipment of the advanced £1000 technology type, and that this was clearly impossible. Schumacher therefore advocated an intermediate technology – a £100 technology (the animal-drawn plough) – which would be affordable to people in the Third World, but which would improve efficiency, reduce drudgery, and help develop and improve the economy.

In the context of construction technology, concrete can be mixed slowly and inefficiently by hand, using a flat wooden board and a shovel (£1 technology). Alternatively, a mechanized concrete batching and mixing plant can be used (£1000 technology). The labour-intensive method is very slow and the quality of the concrete is likely to be inferior. The mechanical mixer is not only expensive, but it may be difficult to maintain owing to lack of local skilled mechanics and the difficulty and cost of obtaining spare parts. Supplies of electrical power or fuel for the mixer may also be unreliable.

An intermediate technology concrete mixer, developed in Ghana, is illustrated in Figure 17.1. It consists of a box lined with thin galvanised metal and fitted with a hinged top, mounted on simple wooden wheels, and fitted with a handle. Sand, aggregate, cement, and water are placed in the box, the lid is fastened and the device is then simply pushed around the site until the concrete is mixed. The device is cheap, simple to maintain and repair, requires no source of power apart from human labour, and produces quite good-quality concrete: laboratory tests indicated cube strengths of approximately 90% of those obtained from a mechanical mixer.

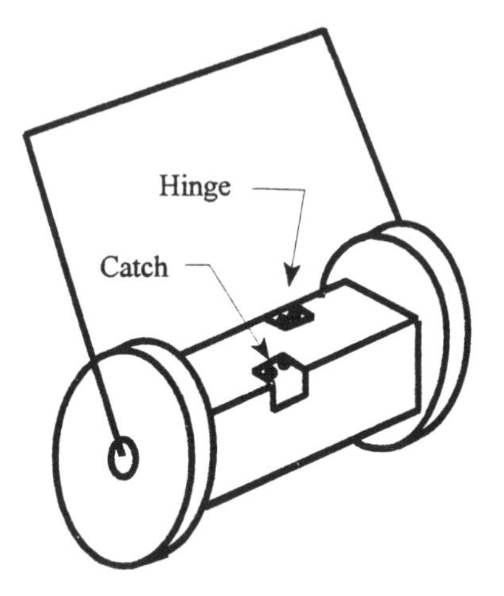

Figure 17.1 Intermediate technology concrete mixer.

There are critics of the use of intermediate technology, including many from developing countries themselves, who point to the rapidly expanding economies of Korea, Malaysia, and Taiwan as examples, and argue that, without access to the most advanced technology available, the Third World will never catch up with the industrialized world. There are numerous examples illustrating the unsustainability of advanced technology: the factories operating at a fraction of their design capacity because of inadequate distribution networks, or lack of maintenance staff or spare parts; the water and wastewater treatment plants that are not functioning because of lack of funds for consumables and spare parts; and the impassable sealed roads, which have not been resurfaced owing to lack of suitable plant or materials.

The project manager must decide on what is or is not appropriate in any given context, and needs to address the questions of fitness of purpose, maintainability, cost, and sustainability. Whether 'local indigenous', 'intermediate', or 'advanced' technology is the most appropriate will depend on the physical, social, cultural and economic environment of the particular developing country in which the project is set.

17.5 Labour-intensive construction

One facet of appropriate construction technology is the use of less plant and equipment, and more labour. This is termed *labour-based* or *labour-intensive construction*. Much of the infrastructure – sewers, aqueducts, canals, railways – in industrialized countries was built by our ancestors using labour-based methods, and it is only fairly recently that extensive use has been made of heavy machinery for construction and building. With the change to machine-based construction, engineers and project managers, in both industrialized and developing countries, became less familiar with labour-based methods. Training for civil engineers has become based on equipment-intensive methods, and this has become the norm throughout most of the world. This situation was encouraged in many developing countries through the provision of aid, particularly bilateral tied aid, by funding agencies, expecting or requiring trade and business as a condition for loans and grants.

In the 1970s it became clear that the plight of the world's rural poor was not improving, partly because the small and isolated infrastructure projects required for these rural areas were not attractive either to funding agencies or to large equipment-based contractors. Research into the potential for labour-based construction revealed that there was a reluctance to adopt such methods because it was felt that the costs could not be accurately predicted, the labour force were unreliable, and that it would be more expensive and more prone to delays than equipment-based construction. In addition to this were the many problems associated with the welfare of large numbers of labourers.

However, the reliance on plant-intensive construction has a number of drawbacks for developing countries, the main one being the high expenditure of foreign exchange, something that few Third World countries can afford. In contrast, labour-based construction actually generates employment and produces an income for those engaged on the project.

It has now been demonstrated that, in certain circumstances, labour-intensive methods can compete with plant-based methods both in terms of technical quality and cost. This can be made possible with good management. Good management of labour-based construction entails ensuring that the morale and motivation of workers remain high at all times by offering incentive payments, providing good training, attending to the welfare of the workers, and ensuring that materials and tools are always available and in good condition. Because labour-based construction projects can often entail work being carried out at a number of small, dispersed and remote sites, good communications between the various sites and good planning and coordination are essential.

Most labour-based construction projects will in fact use a combination of various forms of motive power – human, animal, and machine. The key to efficient construction is selecting the optimum combination for a particular project. Long-distance haulage is probably best done by truck; animals are better and cheaper for shorter distances over steep or rough terrain; and even shorter haulage distances might be most suited to labourers with wheelbarrows or headbaskets. In many construction operations a mix of labour and machine will provide the optimum solution – earthworks may use labour for excavation and a truck for haulage, for example. Balancing these resources to ensure full productivity is one of the tasks of the manager.

Some activities are always more suited to machine-based construction, while others are better carried out by labour-based methods. For example, although a concrete pipe culvert might be installed using machines, brick arch culverts can only be constructed using labour. An asphalt road surface requires equipment, and it would be totally impractical to consider labour-based methods for this form of construction. Earthworks, excavation and quarrying are all well suited to labour-based methods.

The implicit choice of equipment-based construction methods has an influence on the planning and design of the works, and a fair comparison between the financial costs of labour-based and equipment-based methods may be difficult once the design is finalized. Therefore, if labour-based methods are to be given an unprejudiced assessment, alternative designs suited to labour-based construction should be considered. This is sometimes referred to as *design equivalence*. For an even more comprehensive comparison, a full economic appraisal should be carried out, whereby the economic benefits of enhanced employment and the potential for growth and development are included as benefits of the project.

Finally, it is important to note that the availability of labour in rural agricultural areas can vary considerably with the seasons: during planting and harvesting time, full employment and higher wages can be obtained on the farms. It may be necessary to raise project wages at these times in order to ensure continuity of work, although this may have the disadvantage of workers' refusing to return to normal rates of pay later.

It may also have the added drawback to the community of decreased agricultural output. In some cases, it may be necessary to employ migrant workers from another area, and this will entail providing suitable accommodation, normally in camps, which must be provided with water, sanitation, and other essential services. In any case, there is a need to collect data on availability of labour prior to making the decision on whether or not to adopt a labour-intensive approach to the project.

17.6 Community participation

In industrialized countries, the general public may be involved in the sanction and approval stage of projects through public enquiries or public protest, but they are unlikely to be involved in the planning and design of projects, and it is even less conceivable that they would participate in their construction or operation. For projects in developing countries, particularly small projects in rural areas, the concept of 'community' or 'beneficiary' participation is now accepted as being expedient, if not essential, for success.

The participation of the community for whom the project was intended was initially employed to provide assistance with the construction of rural projects, such as the development of water, sanitation, and irrigation schemes. This was primarily to reduce costs and to ensure that the local community would have sufficient expertise to enable them to operate and maintain the schemes once they had been constructed. The idea has since been developed further to encompass identification, planning and even design of projects, the argument being that the community have a much better local knowledge, they know what they want, and of course they are aware of all the social, cultural and religious factors that may affect the design.

For project managers this can entail a great deal of additional work, and it can delay the start and completion of a project. The reported advantages are that it will help to ensure that the project is used, and that it can and will be successfully operated and maintained.

In some cases, such as the improvement of water quality, rural communities may not be aware of the advantages of a project, and community training and education may be necessary. The construction of pilot schemes can be necessary in order to convince villagers of the benefits to be derived from a project and to obtain their views on the design and possible improvements. This all places an extra burden on the project manager, and highlights the importance of communication skills and the ability to work with a variety of different people and organizations.

17.7 Technology transfer

The Third World needs new construction technologies and management expertise in order to develop, but only if those technologies and techniques are appropriate. The long-term benefits acquired when new technology is introduced and used for one short contract will be negligible, and its introduction may possibly even be detrimental to the achievement of long-term goals. However, if there is a real commitment to the ideal of technology transfer in terms of the acquisition of knowledge, skills and equipment, the benefits may be considerable.

Effective technology transfer is more than just education and training. Although training is essential, a much greater understanding of techniques, processes and machinery will be acquired from their actual use than from merely observing or learning about their use. Even more control and mastery will be obtained by owning and being responsible for them. This can be achieved through direct purchase of equipment, entering into a joint venture, becoming a licensee or franchisee for an established process, or entering into some other form of contractual arrangement with an experienced company or organization in the developed world.

Direct purchase of equipment involves the least amount of risk on the part of the vendor and the greatest risk for the purchaser. However, provided that good training, reliable after-sales support, established channels for the supply of spare parts, and sound warranties are included in the package, this can be a quick and effective method of transferring some technologies.

Becoming a licensee may be seen as purchasing intellectual knowledge in addition to purchasing or leasing equipment. It is usually in the interests of the licensor to provide a higher level of support than might be expected from a vendor because the terms of the licence would normally entail payment of a percentage of the value of the work carried out. Hence some of the risk of such a venture will be taken by the licensor.

Joint ventures between an established contractor in an industrialized country and a contractor in a developing country can be a very effective way to transfer technology and encourage the development of the indigenous construction company. There are a variety of different forms of joint venture, all of which require some investment from both the partner in the developed country and the one in the developing country. The investment from the developed country partner is in the form either of cash or of equipment together with technical and management expertise. The developed country partner may invest either cash and/or premises, labour, and locally available equipment. Joint ventures will only be an effective way of transferring technology if both parties are firmly committed to the idea. Very often joint ventures are formed merely to satisfy a local requirement for overseas contractors to enter into joint venture with a local company in order to bid for work in that country. In such cases there is often little or no commitment on either side, the overseas company being involved out of necessity and the local company allowing their foreign partner to make the decisions while they take a back seat – and their share of the profits! Joint ventures set up for short one-off projects are likely to be a less effective means of technology transfer than long-term ventures.

Contractors working in developing countries can help the process by providing training to local staff and subcontractors, and most funding agencies allow for training within the loans or grants awarded for projects. Agencies also provide finance specifically for training and technical assistance, which can be provided through an arrangement known as *twinning*. Twinning is a formal professional relationship between an organization in a developing country and a similar but more mature and experienced organization in another country. Unlike training programmes, these are often long-term arrangements and may involve lengthy visits from key personnel in both organizations to their counterparts in the twinned institution.

17.8 The future

The world's population is still growing at an alarming rate, and much of this growth is in the less developed world. The world urban population is growing even faster, and once again it is in the developing world where the largest and fastest growing cities are to be found. Although infrastructure development is taking place, it is barely keeping pace with the ever-expanding population, all of whom require the basic necessities of food, shelter, water and education. The need is overwhelming. Without civil engineers and people who can effectively manage civil engineering projects this need will never be met. But successful projects in developing countries require managers who recognize the needs and are knowledgeable of the clear and distinctive differences, difficulties, peculiarities and rewards of managing projects in these countries.

Further reading

Cooper, L. (1984) *The twinning of institutions – its use as a technical assistance delivery system.* World Bank Technical Paper 23.

Coukis, Basil and World Bank staff (1983) *Labor-based Constructive Programs*, Oxford University Press.

Institution of Civil Engineers (1981) *Appropriate Technology in Civil Engineering*, Thomas Telford, London.

Smillie, I. (1991) *Mastering the Machine – Poverty, Aid, and Technology*, Intermediate Technology Publications, London.

Vickridge, I. and Boysen, A. (1994) The benefits and constraints of transferring trenchless technology to developing countries. In: *Proceedings of NO-DIG '94*, 11th International Conference of the International Society for Trenchless Technology, ISTT, Copenhagen, pp. B1–B12.

World Bank (1984) *The Construction Industry – Issues and Strategies in Developing Countries*, The World Bank.

World Bank (1992) *World Development Report 1992 – Development and the Environment*, Oxford University Press.

Chapter 18
Engineering Project Management Review

Projects that are technically complex or large in value will normally justify the appointment by the promoter of an individual or team whose sole task is that of project management. To be most effective this appointment should be made early in the process, as soon as the proposed project receives serious consideration. The individual or group should thus become responsible for overall control of the project from its inception through to final commissioning. Whenever possible, they should be appropriately qualified professionals, experienced in the type of project, and able to assume overall responsibility for a full range of necessary tasks. The approach will be most effective if there is continuity of service from inception to completion.

18.1 The roles of the parties

The promoter

The ultimate responsibility for the management of a project lies squarely with the promoter, consequently, any project organization structure should ensure that this ultimate responsibility can be effected. This is not to say that the promoter must be involved in the detailed project management, but it does mean that the machinery must be in place for him to make critical decisions affecting the investment promptly whenever they become necessary.

The project manager must ensure that the promoter organization supports the project team with direction, decision and drive; and that it regularly reviews both objectives and performance.

The project manager

The role of the project manager is to control the evolution and execution of the project on behalf of the promoter. This role will require a degree of executive authority in order to coordinate activities effectively and take responsibility for progress. It will be necessary to define the extent of such delegated authority, and the means by which instructions will be received with regard to those decisions that the project manager is empowered to make.

Ideally the project manager should be involved in the determination of the project objectives and subsequently in the evaluation of the contract strategy. The project manager must therefore drive the project forward and think ahead, delegate routine functions and concentrate on problem areas.

If the project manager is to fulfil the task of control of the realization of the project on behalf of the promoter, decisions taken on engineering matters cannot be divorced from all other factors affecting the investment. Control may only be achieved by regular reappraisal of the project as a whole so that the current situation in the design office, on fabrication, on the supply of materials, and on site may be related to the latest market predictions. If this is done the advantage to be gained, say, from early access to land may be equated with any additional costs in full knowledge of the value of early or timely completion. The continual updating of a simple 'time and money' model of the project originally compiled for appraisal will greatly facilitate effective control during the engineering phase of the project.

18.2 Guidelines for project management

The project management process is briefly summarized below:

(1) The success of a project or contract depends on the management effort expended by the promoter prior to sanction and by both parties prior to award of a contract.

(2) The promoter commits himself to investment in the project on the basis of the appraisal completed prior to sanction. The appraisal must be realistic and identify all risk, uncertainties, potential problem areas and opportunities. Single-figure estimates are misleading and should be supported by figures showing the range of likely outcome of the investment.

The overriding conclusion drawn from recent research is that promoters and all parties involved in construction projects and contracts benefit greatly from reduction in uncertainty prior to their financial commitment. Money spent early buys more than money spent late. Willingness to invest in anticipating risk is a test of a promoter's wish for a successful project.

(3) It is essential that project management ensures that the promoter clearly defines the project objectives together with the ranking of their relative importance. The likelihood of a successful project is greatly improved when all key managers of design, construction and supporting groups are fully informed and committed to these objectives. The project objectives should also be communicated to the other parties involved in project implementation. The dominant considerations must be fitness for purpose of the completed project and safety during both the implementation and operation phases.

Thereafter the normal primary objectives are concerned with cost, time and quality. These are interrelated and may conflict. The fact that the promoter usually does not see any return on his investment until the project is commissioned suggests that timely completion should be a priority.

(4) Engineering projects are normally of short duration and are completed against a demanding time-scale. Adequate staff of the right quality must therefore be appointed and given training in the appropriate techniques and procedures. All staff concerned with contract management must be familiar with the contractual procedures employed.

For both the promoter and contractor, one person in each organization, the project manager, must be ultimately responsible, and be known to be responsible, for the realization of the project.

(5) Although the scope of the project will be agreed at sanction it is probable that conceptual design, which will determine the final layout and size of the functional units, will follow early in the engineering phase. If the conceptual design is rigorously reviewed this provides an opportunity both for cost saving and for ensuring that the proposals meet the promoter's objectives. Particular attention should be given at this stage to subsequent operation and maintenance of the project.

Regular review or project audit of both objectives and achievement should be linked with updating of the project plan.

(6) Effective control of the project will only be achieved through continual planning and replanning. Management effort should be concentrated on the present and the future: time devoted to the reporting and collection of historical data should be kept to a minimum.

In his planning the promoter and his project manager must take a broad view of the project and aim to coordinate design, implementation, commissioning and subsequent operation and maintenance. Interaction of contractors, access, statutory requirements and public relations must all be considered.

A contractor will plan in detail and aim to achieve continuous and efficient deployment of his resources. Owing to the likelihood of change, the contractor's programme should be flexible and subject to constant review by the project manager.

(7) A strategic project plan should show clearly the financial consequences of alternative courses of action and of indecision. It is therefore convenient to develop the plan as a time-and-money model of the project, which will react realistically to changes in timing, method, content and cost of work. Realism is largely dependent on the correct definition and allocation of costs and revenues as either fixed, time-related or quantity-proportional charges. Time-related costs are significant in all types of construction work and predominate in many civil engineering projects. Adherence to the programmed time schedule for the work will therefore also control both cost and investment.

(8) Time lost at the beginning of a project can rarely be recovered: particular attention must therefore be given to the start-up of the project. Similarly, sufficient time must be allowed for mobilization by each contractor.

(9) Consideration of alternative contract strategies will frequently focus attention on deficiency of information and on the problems that will hinder achievement of the project objectives. Selection of an appropriate contract strategy at an early stage of project implementation is perhaps the most important single activity of the project management team.

(10) Appointment of a contractor on the sole criterion of lowest bid price will not necessarily lead to a harmonious contractual relationship. The lowest tender may not ultimately produce the lowest contract price.

Both parties are making their commitment at this point and should be fully aware of both the promoter's objectives and the contractual responsibilities. Selective tendering followed by rigorous bid appraisal, including study of the contractor's programme and resource allocation, will do much to ensure that the contractor has not misjudged the job and that his price is realistic. The production of his own operational type of cost estimate will greatly aid the project manager in this appraisal.

The promoter must check that all his obligations can be honoured before award of the contract.

(11) Throughout the implementation period of the project the

promoter will inspect and approve the quality of workmanship of contractors and manufacturers, unless Quality Assurance Certificated. Again, an adequate number of staff with relevant experience must be employed. Prior definition and agreement of acceptable standards is essential, and all parties should be aware of tolerances. There is a tendency for design engineers to specify unnecessarily high standards, the achievement of which may prove difficult and/or expensive. The desired quality of workmanship must always be considered in relation to the promoter's other prime objective, usually timely completion and economical cost.

(12) Promoters frequently underestimate both the extent and consequence of change. The project manager should rigorously assess the cost and benefit of all design changes before they are implemented. Priority should be given to timely completion of the project.

The better organized the contractor, the more likely it is that he is working to a tight, well-resourced programme. The disruptive effect of variation may therefore be serious. Modifications to manufacturing plant are sometimes best implemented during some future shutdown of the plant for maintenance.

(13) Involvement in prolonged bargaining over claims is a sign of failure. Evaluate and agree payment for variations and claims as the job progresses. The valuation should be based on prices, resource output and efficiencies similar to those incorporated in the contractor's tender.

(14) Total quality management (TQM) can assist in the setting and achievement of project objectives. Properly practised, TQM requires precision of communication, which can be considered as its greatest value to project management. Care must be taken to ensure that the adoption of a quality system does not result in rigid adherence to unnecessarily demanding specifications. Neither must the system inhibit the flexibility and judgement required for the management of the uncertainties associated with the project.

(15) In the context of project management, the quality of performance of the project is greatly dependent on the quality of project staff. Projects are managed by people who are continuously directing and communicating with other people. Great attention must be paid to the selection and motivation of staff. Personality and ability to think ahead are as important as technical know-how.

Management is concerned with the setting and achievement of realistic objectives for the project. This will demand effort; it will not happen as a matter of course; and it will require the dedication and motivation of people. The provision and training of an adequate management team is therefore an essential prerequisite for a successful job, for it is their drive and judgement, their ability to persuade and lead, which will ensure that the project objectives are achieved.

18.3 Project management: the way ahead

It is always difficult to prejudge the evolution and development of existing systems, particularly in the more subjective management area. The further into the future the target, the less can be seen, but prediction has to be attempted if progress is to be made. Prediction is based on the analysis of the existing situation and the related historic trend that is used as a mechanism to project into the future. This technique may be acceptable for many purely technological processes and for short-term prediction but it is limited in its ability to encompass the wider picture.

Nevertheless, there are a number of developments in project management theory and practice that would appear to have a relatively high probability of occurrence. The current developed world market recession has placed increasing demands upon the need to manage and to deliver projects in an effective way. This has inspired a plethora of management approaches, some new, some revised versions of existing ideas, which are seen to be either compatible with, or in competition with, project management. In particular, several approaches seem to have become widely established; value management, *kaizen*, total quality management, waste management and just-in-time management.

Project management relies upon good management practice but has the overall goal of project completion in terms of the project objectives as its prime objective. The techniques mentioned above are also aimed at making improvements in management practice and hence increasing the effectiveness of projects, but achieve this aim by concentrating on non-project parameters. The most successful users of these individual techniques are organizations not currently using project management for their internal and external projects, often from a manufacturing or production background. Therefore it is likely that project management will, over the next few years, subsume these non-project management approaches, which is likely to involve some changes in approach and terminology but will result in improved project management procedures integrating discrete business functions to enhance the effectiveness of decision making. Success or failure in this area is likely to determine the longevity of project management as a discrete discipline.

Further reading

Gaisford, R.W. (1986) Project management in the North Sea. *International Journal of Project Management*, **4** (1), 5–12.

Perry, J.G. and Thompson, P.A. (1992) *Engineering Construction Risks – A Guide to Project Risk Analaysis and Risk Management*, Thomas Telford, London.

Thompson, P.A. (1991) The client role in project management. *International Journal of Project Management*, **9** (2), 90–92.

Wearne, S.H. (ed.) (1989) *Control of Engineering Projects*, 2nd edn., Thomas Telford, London.

Suggested Answers to Exercises in Chapter 10

10.1 New housing estate

The project consists of a small development of a new housing estate, familiar to many urban or suburban sites. The project includes four blocks of low-rise flats, a block of five-storey flats and a shops and maisonettes complex, with support services.

The key issue is to identify where most of the work is required. In this case the five-storey block absorbs the majority of the site man-hours, and unless careful assumptions are made regarding the gang sizes and the overlapping of the different stages of construction, it would not be possible to finish within the time allowed. Therefore, despite having blocks A and B as a priority, as soon as the drain has been diverted work must start on the five-storey block.

There is no single correct solution, as the precise answer depends upon the assumptions made. The model solution suggested is shown in Figure S1. Assumptions made include having two excavation teams of 8 men: the first team commences on block A and then goes to the five-storey block; the second team commences on block B and then also to the five-storey block. The concrete frame for the five-storey block is assumed to have 16 men, the brickwork 12 men and the finishers 20 men. It is further assumed that brickwork could overlap the frame by 14 weeks and that finishing could overlap brickwork by 15 weeks.

This results in a base demand for 12 bricklayers on site, rising to a plateau of 24 bricklayers for 20 weeks before falling back to 12 again. The logic is shown by dotted lines, and important constraints or key dates are clearly marked. The space within the bars has been used for figures of output. A histogram of the demand for bricklayers and for total labour has been plotted directly under the bars at the bottom edge of the programme.

This solution completes in 78 weeks, allowing 4 weeks float to compensate for an optimistic programme, and has a maximum number of

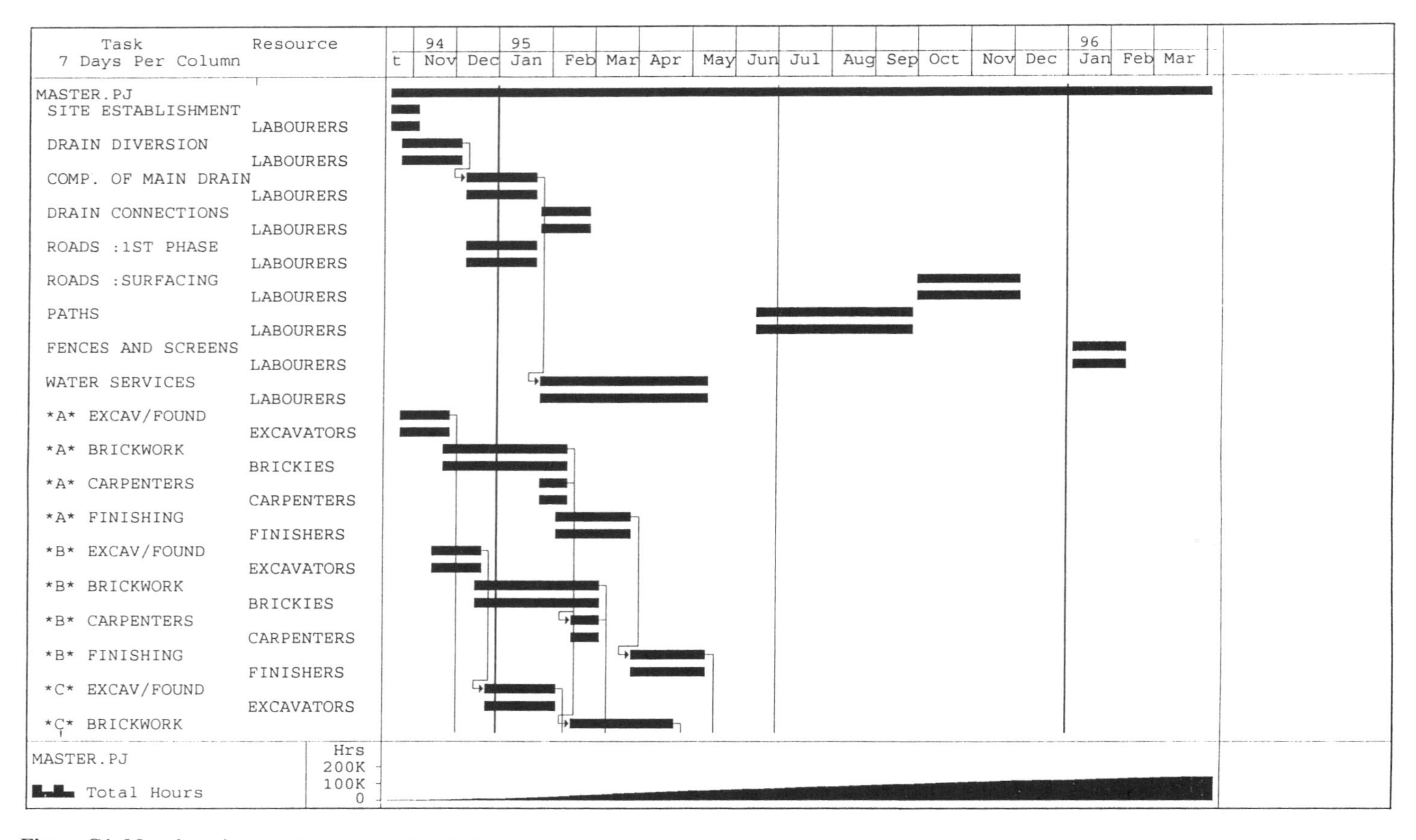

Figure S1 New housing estate: suggested solution.

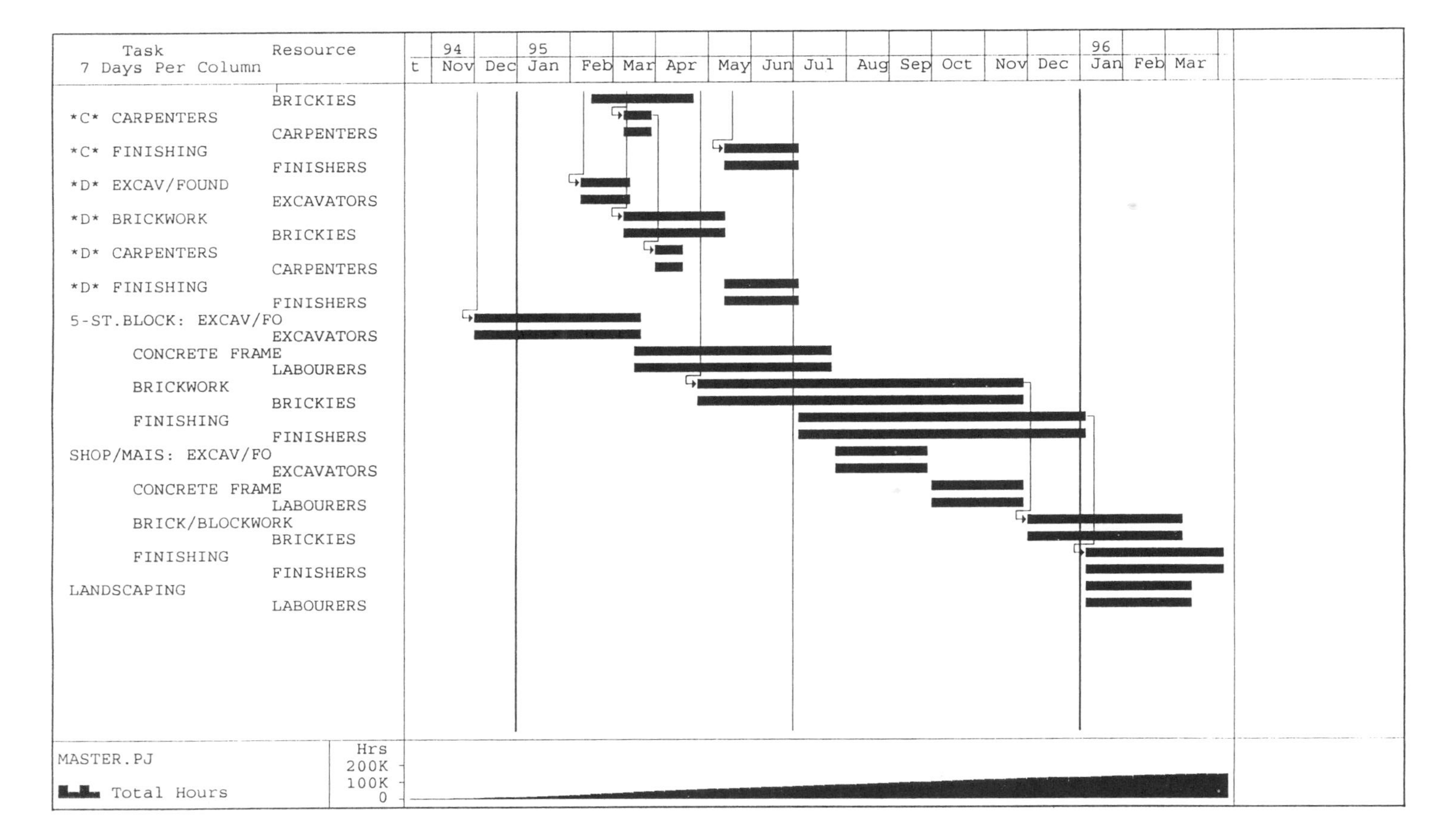

Figure S1 *(Contd).*

people on site of 65. Keeping the total on site low is important in practice, as accommodation and equipment has to be provided for all workers. There is a conflict between smoothing the bricklayers and a smooth demand for total labour, and computer-based methods are more efficient in resolving complex resourees conflicts. With this solution flats A and B are ready by week 34.

It is important to note that for A and B to be habitable then assumptions have to be made about the proportions of work for the drain connections, paths, fences and screens, water services and landscaping that will have to be completed by week 34 also.

10.2 Pipeline

The time–location diagram is particularly suitable for cross-country jobs such as pipelaying, where the performance of individual activities will be greatly affected by their location and the various physical conditions encountered. In this exercise the assumption is that one stringing gang and four separate pipelaying gangs are to be employed. The remaining problem is to decide whether a single bridging gang can cope with all restrictions placed on the programme.

The programme has to adopt a trial and error approach to producing the diagram, to balance the needs for bridging with the productivity of the gangs. The time-location diagram, shown as Figure S2, shows a possible solution for one bridging gang, one stringing gang and four pipelaying gangs. Pipelaying gang 1 starts at 0 km, gang 2 at 9 km, gang 3 at 12 km and gang 4 at 16 km. It can be seen from the slopes of the progress lines, falling from 300 m/week/gang in normal ground to 75 m/week/gang in rock, that there is a need for two gangs to tackle the rock between 10 km and 13 km. The bridging gang move to keep ahead of each of the pipelaying gangs approaching an obstruction and hence move from the culvert at 2 km to the culvert at 18 km, to the thrust bore at 13 km, to the river crossing at 7 km, and finally to the railway crossing at 22 km.

Pipes are available at a maximum rate of 1000 m/week. Remembering the constraint to keep at least one week ahead of the pipelayers, a sequence for offloading from suppliers' lorries, storing and stringing out is shown on a weekly basis.

10.3 Industrial project

(a) The precedence diagram for this exercise is shown in Figure S3. The activities are represented by egg-shaped nodes and the interrelationships

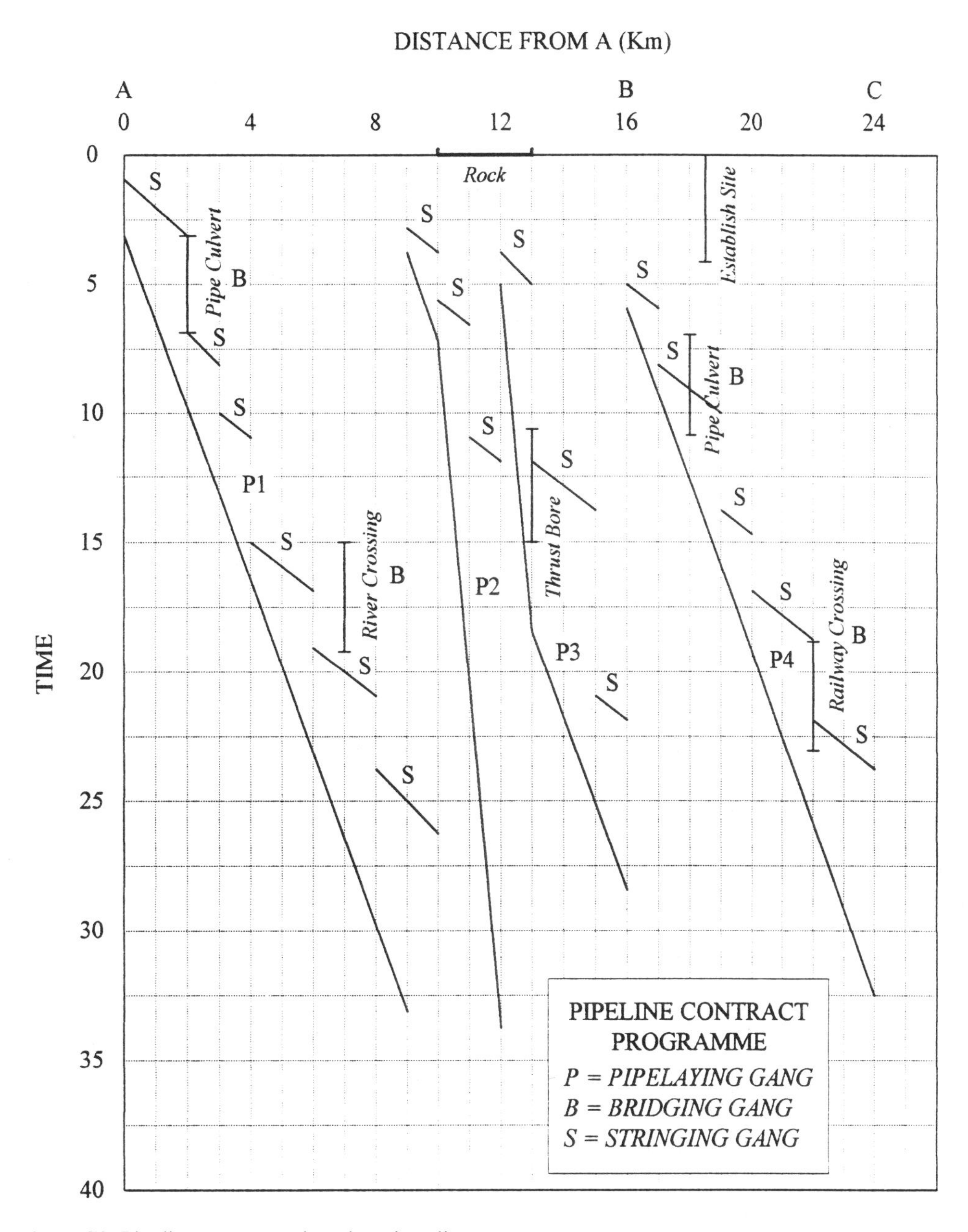

Figure S2 Pipeline contract: time–location diagram.

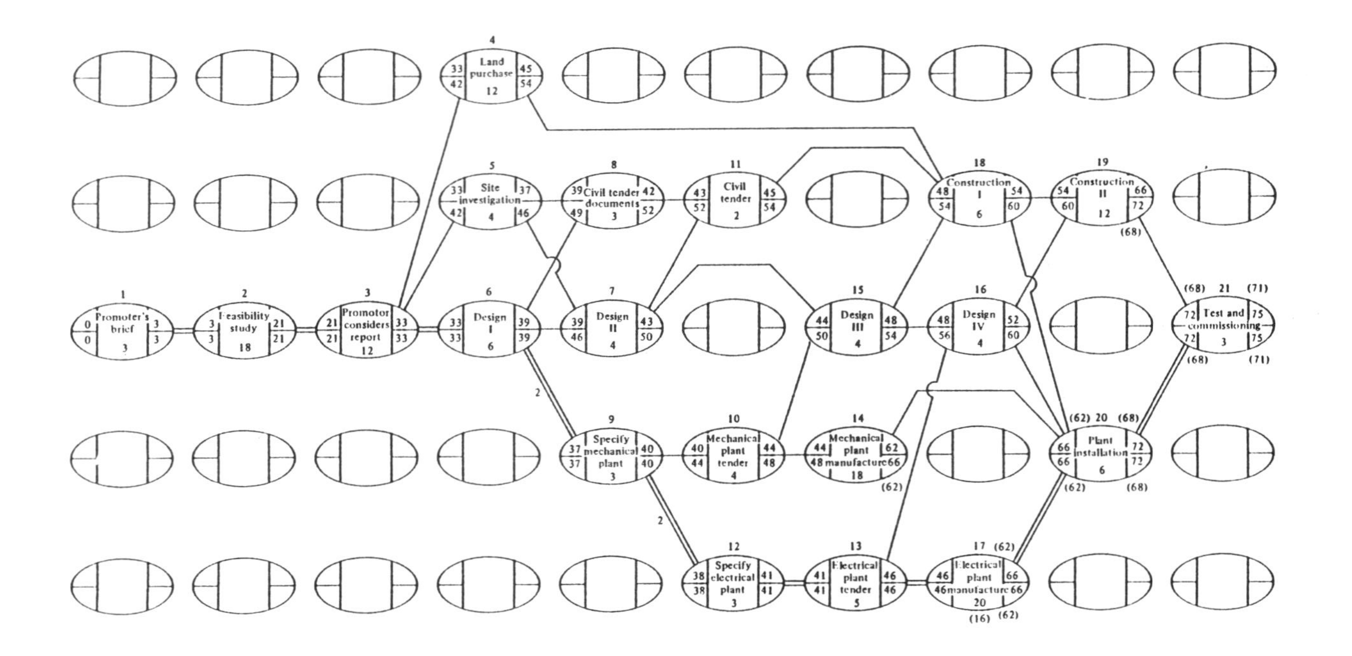

Figure S3 Industrial project: Outline project network (precedence diagram). All durations are expressed in months.

between activities by lines known as *dependences*. The diagram is constructed and the forward and backward passes undertaken in exactly the same way as the worked example in the chapter. Minimum duration is 75 months and the critical path goes through activities 1–2–3–6–9–12–13–16–20–21.

Note: It is unlikely that you will get the most useful diagram at the first attempt. When redrawing a critical path diagram try to minimize the crossing of logical dependences, although often some crossing is unavoidable, but more importantly think about the role of the plan in communicating to the project team. In Figure S3 each level or row of the diagram represents a particular responsibility: top level, land purchase (legal department); second level, site investigation and civil (civil engineering); main level, project activities and design; fourth level, mechanical department; and fifth level, electrical department. This is one way of presenting the information, but there are many other ways that might be particularly useful in a given situation.

The float associated with activity 16 is as follows. *Total float*, the difference between its earliest and latest starts or finishes, is 8 months; *free float*, the minimum difference between the earliest finish time of that activity and the earliest start time of a succeeding activity, is 2 months.

(b) The option to spend more money to save time is frequently encountered in practical project management. In this simple example the combinations of A and C represent the existing situation and A and D, B and C, and B and D require investigation. However, activity 14 is not on the critical path, and therefore A and D would not be beneficial and can be ignored.

Take Activity 17 first and substitute B for A. Recalculate the remainder of the forward pass to show an early finish of 71 months. Calculating the backward pass provides two critical paths: 1–2–3–6–9–12–13–16–20–21 and 1–2–3–6–9–10–14–20–21. The extra cost of substituting B for A is £156 000 but the profit earned would be (75–71) 4 × £50 000 = £200 000, a gain of £44 000. Therefore option B should be adopted for activity 17.

Activity 14 is now on one of the critical paths. However, as there are two paths, a separate reduction in the duration of activity 14 would not reduce the project duration, and hence would not be cost effective; option C should be retained.

Therefore options B and C are the recommended choice.

10.4 Bridge

(a) Taking a step closer to the real world, no logic dependences are indicated, and the planner has to use judgement to construct the network. There are a number of possible networks, but it is recommended that the diagram should be kept as free and unconstrained as possible. It is suggested that 'set up site' would be the start activity, and this would then link to all the excavation activities. The piledriving for the right pier should be inserted after 'excavation' but before 'foundations'. All other 'excavation' activities are followed by 'foundations'. 'Foundations' are followed by the next stage of the 'Concrete' process. Next, beams can be placed, but remember in the logic that for a beam to be placed it must have the supports at both sides completed. Activity 18 'clear site' is suggested as the finish activity.

No figure is included, as there are a number of viable solutions. One of the optimal solutions gives completion in week 30 with two critical paths, both going through activity 6, the right pier pile driving.

(b) The effects of reducing the duration of pile driving to 5 weeks will vary depending upon the network drawn for part (a). Nevertheless, the overall effect on most networks is to reduce the completion time to 29 weeks with two different critical paths, one going through the north, and one the south, abutment activities.

(c) The resources are heavily constrained. One Excavation team, one concrete team for foundations, one concrete team for abutments and piers and one crane team severely limits the activities that can operate in parallel.

The question is slightly unfair, as this type of resource scheduling is difficult to undertake without the assistance of a computer. Many people find the use of resourced bar charts of assistance, but there is no easy way. In this case the obvious action is to employ resources away from the right pier such that the extra work required for the piledriving does not delay or disrupt the use of any resource.

If this is achieved, a completion time for a 'resource-constrained' project of 46 weeks can be achieved. However, any solution under 50 weeks represents a reasonable attempt at the exercise.

Index